Discrete Stochastic Models and Applications for Reliability Engineering and Statistical Quality Control

Discrete stochastic models are tools that allow us to understand, control, and optimize engineering systems and processes. This book provides real-life examples and illustrations of models in reliability engineering and statistical quality control and establishes a connection between the theoretical framework and their engineering applications.

This book describes discrete stochastic models along with real-life examples and explores not only well-known models, but also comparatively lesser-known ones. It includes definitions, concepts, and methods with a clear understanding of their use in reliability engineering and statistical quality control fields. Also covered are the recent advances and established connections between the theoretical framework of discrete stochastic models and their engineering applications.

This book is an ideal reference for researchers in academia and graduate students working in the fields of operations research, reliability engineering, quality control, and probability and statistics

Discrete Stochastic Models and Applications for Reliability Engineering and Statistical Quality Control

Serkan Eryilmaz

CRC Press
Taylor & Francis Group
Boca Raton London New York

CRC Press is an imprint of the
Taylor & Francis Group, an **informa** business

First edition published 2023
by CRC Press
6000 Broken Sound Parkway NW, Suite 300, Boca Raton, FL 33487-2742

and by CRC Press
4 Park Square, Milton Park, Abingdon, Oxon, OX14 4RN

CRC Press is an imprint of Taylor & Francis Group, LLC

ISBN: 978-0-367-74983-5 (hbk)
ISBN: 978-0-367-74984-2 (pbk)
ISBN: 978-1-003-16063-2 (ebk)

DOI: 10.1201/9781003160632

Typeset in Times New Roman
by KnowledgeWorks Global Ltd.

Publisher's note: This book has been prepared from camera-ready copy provided by the authors.

Dedication

To my beloved parents Nurhan and Necati Eryilmaz,
and my lovely daughter Buse Ceylin Eryilmaz

Contents

Preface

It has become increasingly apparent that many problems involving uncertainty in various fields of science provide important examples to discrete stochastic models. There are many real-life engineering systems and processes which may be described by discrete stochastic variables. The number of failed wind turbines during a specified period of time in a wind farm, the total number of collected subsamples until an out-of-control signal in a specific production process, the number of failed components in a technical system at the time when the system fails, and the number of arrivals to a certain service station can be all modeled by a discrete stochastic variable. Further, the lifetimes of many real-life systems/components are defined as the number of cycles until their failure. For example, the lifetime of car tires are measured in terms of the number of kilometers or the number of days; the lifetime of a switch is measured by the number of times it is switched on/off.

As it is well known, the most important probabilistic tool for understanding such a stochastic variable is its probability distribution. Thus, the development of suitable discrete stochastic models is of special importance for proper analysis of the variable under concern, and hence, the corresponding real-life phenomenon or system.

The purpose of this book is to present the well-known and lesser-known discrete stochastic models with an emphasis to their applications in reliability engineering and quality control. It is not intended to present the complete theory of the topics mentioned. Rather, a lucid description of the models is presented with simple real-life examples. In the proofs and explanations, clarity and simplicity are favored. Extensive subchapters are presented on specific research fields of reliability engineering and statistical quality/process control.

This book helps the reader develop the understanding and intuition necessary to apply discrete stochastic models in reliability engineering and operations research. It is organized into eight chapters. Chapter 1 provides an overview and discussion of the discrete stochastic modeling. In Chapter 2, well-known and lesser-known (but with many real-life applications) discrete probability distributions are reviewed. Particular discrete stochastic models are described through Chapters 3–7 with simple and clear examples. Chapter 8 contains extensive applications concerned with important fields including system reliability models, shock models, age replacement policies, and statistical quality control.

I wish to express my gratitude to all who have supported me in one way or another throughout my academic journey, especially my Ph.D. supervisor Professor Ismihan Bairamov (Bayramoglu) who has greatly contributed to my learning of probability theory and stochastic modeling. I extend my appreciation to my former Ph.D. student Dr. Cihangir Kan and Dr. Murat Ozkut for their great assistance in formatting the book. My sincere thanks goes to my dear friends and colleagues, Professor Yılser Devrim and Professor Fatih Tank, who have always been there for me. I would also like to extend my deepest appreciation to my beloved daughter Buse Ceylin Eryilmaz for being on her best behavior, and always cheering me up.

I would like to thank the readers in advance for informing me of any errors. Any corrections to errors found in the book will be posted on the website: www.serkaneryilmaz.com.

Serkan ERYILMAZ
Ankara, Turkey
December 2021

Author Biography

Serkan Eryilmaz completed his Ph.D. at Ankara University (Turkey) in 2002. He is currently working as a professor at the Department of Industrial Engineering and serving as vice president of research at Atilim University, Ankara. He is the present co-chair of the System Reliability Technical Committee of the European Safety and Reliability Association. He currently serves as a member of the Editorial Board of Reliability Engineering and System Safety and Area Editor of the IISE Transactions-Quality & Reliability Engineering. His research interests include reliability engineering, applied probability, stochastic modeling, and renewable energy systems. He is the author of more than 180 publications. He was honored by TUBITAK (Scientific and Technological Research Council of Turkey) Science Incentive Award due to his highly qualified studies on system reliability in the field of stochastic models/processes in operations research

1 An Overview of Discrete Stochastic Modeling

Discrete stochastic structures widely appear in real life. Discrete random variables and their characteristics are useful in modeling and analyzing various structures of a discrete nature. A random variable that may take on at most a countable number of possible values is called discrete. Some examples for discrete random variables are given below:

* The lifetime of a jet tire which is measured in terms of number of flights.
* The number of customers waiting in a queue corresponding to a certain service station.
* The lifetime of the switch which is measured by the number of times it is switched on/off.
* The number of component failures in a certain technical system.
* The number of available (working) wind turbines within a certain wind farm.
* The demand values arrived in a certain inventory management system.

1.1 DISCRETE RANDOM VARIABLE

For a discrete random variable X that may take the values $x_1, x_2, \ldots$, the probability mass function $p(x)$ of X is defined by

$$p(x) = P\{X = x\},$$

for $x = x_1, x_2, \ldots$, where $p(x_i) > 0, i = 1, 2, \ldots$, and

$$\sum_{i=1}^{\infty} p(x_i) = 1.$$

If $F(x) = P\{X \leq x\}$ denotes the cumulative distribution function of X, then we have

$$p(x) = F(x) - F(x-1).$$

The kth moment of X can be computed from

$$E(X^k) = \sum_{i=1}^{\infty} x_i^k p(x_i).$$

DOI: 10.1201/9781003160632-1

In particular, the expected value and variance of the random variable X are

$$E(X) = \sum_{i=1}^{\infty} x_i p(x_i),$$

and

$$Var(X) = E(X - E(X))^2 = E(X^2) - [E(X)]^2.$$

The expected value of a function $h(X)$ is computed from

$$E(h(X)) = \sum_{i=1}^{\infty} h(x_i)p(x_i).$$

The probability generating function of the random variable X is

$$E(z^X) = \sum_{i=1}^{\infty} z^{x_i} p(x_i).$$

Example 1.1. Consider a power system that consists of three independent power units, namely, unit A, unit B, and unit C. The generation capacities of the units are given respectively by 20 MW, 10 MW, and 30 MW. The availability values of the units are $p_A = 0.95, p_B = 0.97$, and $p_C = 0.93$. If the power generated by the unit i is denoted by X_i, $i = A, B, C$, then

$$\begin{aligned} P\{X_A = 20\} = 0.95 \ , \ P\{X_A = 0\} = 0.05, \\ P\{X_B = 10\} = 0.97 \ , \ P\{X_B = 0\} = 0.03, \\ P\{X_C = 30\} = 0.93 \ , \ P\{X_C = 0\} = 0.07. \end{aligned}$$

At any instant of time, the capacity of the whole power system can be 0, 10, 20, 30, 40, 50, or 60 MW. The states of the units corresponding to these capacities are presented in Table 1.1.

Table 1.1
The States of the Units and Total Capacity of the System

Unit A	Unit B	Unit C	Total capacity (MW)
1	1	1	60
1	1	0	30
1	0	1	50
0	1	1	40
1	0	0	20
0	1	0	10
0	0	1	30
0	0	0	0

Because at any instant of time, the plants are either available or unavailable, state probabilities corresponding to the power system are computed as

$$\begin{aligned}
P\{X_S = 0\} &= (1-p_A)(1-p_B)(1-p_C) = 0.0001\\
P\{X_S = 10\} &= (1-p_A)p_B(1-p_C) = 0.0034\\
P\{X_S = 20\} &= p_A(1-p_B)(1-p_C) = 0.002\\
P\{X_S = 30\} &= p_A p_B(1-p_C) + (1-p_A)(1-p_B)p_C = 0.0659\\
P\{X_S = 40\} &= (1-p_A)p_B p_C = 0.0451\\
P\{X_S = 50\} &= p_A(1-p_B)p_C = 0.0265\\
P\{X_S = 60\} &= p_A p_B p_C = 0.857,
\end{aligned}$$

where X_S denotes the state of the power system as a total capacity. The expected power output of the system is then computed as

$$E(X_S) = 56.6 \text{ MW.} \quad \blacksquare$$

Some systems or processes can be described by more than one random variable and a statistical dependence may exist between the random variables. For example, in a queueing system with queue-length-dependent service times, the service times of customers depend on the queue length (see, e.g. Choi at al. 1999). A system may consists of two or more dependent components (see, e.g. Oliveira et al. 2019). The joint probability mass function of two discrete valued random variables X and Y is defined by

$$p(x,y) = P\{X = x, Y = y\}$$

for $x = x_1, x_2, \ldots,$ and $y = y_1, y_2, \ldots,$ and it satisfies

$$\sum_{i=1}^{\infty}\sum_{j=1}^{\infty} p(x_i, y_j) = 1.$$

The function $p(x,y)$ provides information about joint behavior of the random variables X and Y. The joint probability mass function of X and Y can also be represented as

$$\begin{aligned}
P\{X = x, Y = y\} = P\{X > x, Y > y\} &+ P\{X > x+1, Y > y+1\}\\
&- P\{X > x, Y > y+1\} - P\{X > x+1, Y > y\}.
\end{aligned}$$

The joint probability generating function of the random variables X and Y can be computed from

$$E(z_1^X z_2^Y) = \sum_{i=1}^{\infty}\sum_{j=1}^{\infty} z_1^{x_i} z_2^{y_j} p(x_i, y_j).$$

For $P\{Y = y\} > 0$, the conditional probability mass function and expected value of X given $Y = y$ can be respectively computed from

$$P\{X = x \mid Y = y\} = \frac{P\{X = x, Y = y\}}{P\{Y = y\}},$$

and

$$E(X \mid Y = y) = \sum_{i=1}^{\infty} x_i P\{X = x_i \mid Y = y\}.$$

We also have

$$E(X) = \sum_{j=1}^{\infty} E(X \mid Y = y_j) P\{Y = y_j\}.$$

The expected value of the sum of random variables is equal to the sum of the expected values of random variables. That is,

$$E(X+Y) = E(X) + E(Y).$$

The random variables X and Y are independent if for all x, y,

$$P\{X = x, Y = y\} = P\{X = x\} P\{Y = y\}.$$

Under the assumption of independence of X and Y,

$$Var(X+Y) = Var(X) + Var(Y).$$

The probability mass function of the random variable $X+Y$ can be computed from

$$P\{X+Y = z\} = \sum\sum_{(x,y):x+y=z} P\{X = x, Y = y\}.$$

If the random variables X and Y are independent, then

$$P\{X+Y = z\} = \sum_{j=1}^{\infty} P\{X = z - y_j\} P\{Y = y_j\}.$$

For two random variables X and Y, the probability that X is less than Y can be computed from

$$P\{X < Y\} = \sum\sum_{(x,y):x<y} P\{X = x, Y = y\}.$$

If the random variables X and Y are independent, then

$$P\{X < Y\} = \sum_{j=1}^{\infty} P\{X < y_j\} P\{Y = y_j\}.$$

In the literature, the probability $P\{X < Y\}$ is usually called stress-strength reliability. (An et al. 2008) studied such a probability via universal generating functions.

Example 1.2. Consider a single unit system that is subject to shocks over time. Under the shock model that has been defined and studied by (Eryilmaz 2015), let c_1 and c_2 be two critical levels such that $c_1 < c_2$. A shock with a magnitude between c_1 and c_2 has a partial damage on the system, and the system transits into a lower partially working state upon the occurrence of each shock in (c_1, c_2). That is, after each shock with magnitude between c_1 and c_2, the system state is decreasing by one unit. A shock with a magnitude above c_2 has a catastrophic effect on the system and it causes a complete failure. Let Y_i be a random variable that denotes the magnitude of the ith shock. Let us define

$$\begin{aligned} p_1 &= P\{Y_i \le c_1\}, \\ p_2 &= P\{Y_i \in (c_1, c_2]\}, \\ p_3 &= P\{Y_i > c_2\}, \end{aligned}$$

for $i \ge 1$ and $p_1 + p_2 + p_3 = 1$. Let W be a random variable that denotes the number of shocks until the first shock above c_2 and Z be the number of shocks until the first shock between c_1 and c_2. Then, as shown by (Eryilmaz 2015), the joint probability mass function of Z and W is

$$P\{Z = z, W = w\} = \begin{cases} p_1^{z-1}(p_1+p_2)^{w-z-1}p_2p_3, & \text{if } z < w \\ p_1^{w-1}(1-p_2)^{z-w-1}p_2p_3, & \text{if } z > w. \end{cases}$$

The probability that a shock in $(c_1, c_2]$ arrives before a shock above c_2 is computed as

$$\begin{aligned} P\{Z < W\} &= \sum_{z<w}\sum p_1^{z-1}(p_1+p_2)^{w-z-1}p_2p_3 \\ &= \sum_{z=1}^{\infty}\sum_{w=z+1}^{\infty} p_1^{z-1}(p_1+p_2)^{w-z-1}p_2p_3 \\ &= \frac{p_2}{1-p_1}. \end{aligned}$$

The marginal probability mass function of Z can be computed from

$$\begin{aligned} P\{Z = z\} &= \sum_{w=1}^{z-1} P\{Z = z, W = w\} + \sum_{w=z+1}^{\infty} P\{Z = z, W = w\} \\ &= \sum_{w=1}^{z-1} p_1^{w-1}(1-p_2)^{z-w-1}p_2p_3 + \sum_{w=z+1}^{\infty} p_1^{z-1}(p_1+p_2)^{w-z-1}p_2p_3 \\ &= p_2(1-p_2)^{z-1}, \end{aligned}$$

for $z = 1, 2, \ldots$. Next, let us compute the conditional expected value of W given $Z = z$, i.e. $E(W \mid Z = z)$.

$$\begin{aligned} E(W \mid Z = z) &= \frac{1}{P\{Z = z\}}\sum_{w} wP\{Z = z, W = w\} \\ &= \frac{1}{p_2(1-p_2)^{z-1}}\left[\sum_{w=1}^{z-1} wp_1^{w-1}(1-p_2)^{z-w-1}p_2p_3 \right. \\ &\left. + \sum_{w=z+1}^{\infty} wp_1^{z-1}(p_1+p_2)^{w-z-1}p_2p_3\right]. \end{aligned}$$

After some simple algebraic manipulations, we obtain

$$E(W \mid Z = z) = \frac{(1-p_2)^z + (1-p_1)p_1^{z-1}}{p_3(1-p_2)^{z-1}}.$$

Example 1.3. Consider the following bivariate geometric distribution which has been studied by (Phatak and Sreehari 1981) and (Krishna and Pramendra 2009):

$$P\{X = x, Y = y\} = \binom{x+y}{x} p_0^x p_1^y p_2,$$

for $x, y = 0, 1, 2, \ldots; 0 < p_0, p_1 < 1$, and $p_2 = 1 - p_0 - p_1$. It can be shown that

$$P\{X = x\} = \frac{p_2}{1-p_1}\left(\frac{p_0}{1-p_1}\right)^x,$$

for $x = 0, 1, \ldots$. The conditional distribution of Y given $X = x$ is

$$P\{Y = y \mid X = x\} = \binom{x+y}{x} p_1^y (1-p_1)^{x+1},$$

for $y = 0, 1, \ldots$. The joint probability generating function of X and Y is

$$\begin{aligned} E(z_1^X z_2^Y) &= \sum_{x=0}^{\infty}\sum_{y=0}^{\infty} z_1^x z_2^y \binom{x+y}{x} p_0^x p_1^y p_2 \\ &= \frac{p_2}{1 - p_0 z_1 - p_1 z_2}. \end{aligned}$$

1.2 CONDITIONING

The method of conditioning is very useful in most of the discrete problem setups. It is actually based on the well-known total probability law. Consider a sequence $X_1, X_2, \ldots, X_n$ of trials with two possible outcomes as either "1" (success) or "0" (failure). Let R_n be the total number of success runs in $X_1, X_2, \ldots, X_n$. A run may be defined as an uninterrupted subsequence of same kind of elements. For example, in a binary sequence of $n = 11$ trials 11000101110, we have three success runs which are given by "11", "1", and "111". If S_n denotes the total number of successes in $X_1, X_2, \ldots, X_n$, then, the probability mass function of the random variable R_n can be obtained by conditioning on S_n :

$$P\{R_n = r\} = \sum_l P\{R_n = r \mid S_n = l\} P\{S_n = l\}. \tag{1.1}$$

Example 1.4. Consider a conventional power system that consists of a certain number of units and each may have different capacities. Let c be the number of units within the conventional system. Assume that the total number of states for the entire conventional system is s. For example, consider the system given in Example 1.1.

Then, $c = 3$ and $s = 7$. Suppose that a wind farm that consists of n wind turbines is incorporated to a conventional generation system. Let m be the number of states corresponding to the output power of a single wind turbine. Then, the probability that the combined system produces power $b(j)$ is

$$\begin{aligned} &P\{P_{sys} = b(j)\} \\ &= \sum_{\substack{\omega=1 \\ k\cdot h(l)+g(\omega)=b(j)}}^{s} \sum_{k=0}^{n} \sum_{l=1}^{m} \binom{n}{k} p^k (1-p)^{n-k} P\{P_c = g(\omega)\} P\{P_{WT} = h(l)\}, \end{aligned} \tag{1.2}$$

for $j = 1, 2, ..., m_{sys}$, where P_{sys} is the power generated by the combined system, P_c is the power generated by the conventional system, P_{WT} is the power generated by the wind turbine, m_{sys} is the total number of states for the combined system, $\{g(1), g(2), ..., g(s)\}$ is the set of possible capacities of the conventional system corresponding to the states $\{1, 2, ..., s\}$, $\{h(1), h(2), ..., h(m)\}$ is the set of possible capacities of the wind turbine corresponding to the states $\{1, 2, ..., m\}$, and $b(1), b(2), ..., b(m_{sys})$ are the possible capacities of the combined system. The parameter p in (1.2) represents the availability/reliability of a single wind turbine.

The equation (1.2) is a direct application of the conditioning on three independent random variables, namely, the number of available (working) wind turbines, the power generated by the conventional system, and the power generated by the wind turbine.

Example 1.5. Let N be a random variable with probability mass function $P\{N = n\}, n = 1, 2, ...$ and $Y_1, Y_2, ...$ be a sequence of independent and identically distributed random variables. Let's define a compound random variable $S = \sum_{i=1}^{N} Y_i$. If the sequence of random variables $Y_1, Y_2, ...$ has common discrete distribution and independent of N, then by conditioning on N,

$$\begin{aligned} P\{S = s\} &= \sum_{n=1}^{\infty} P\{S = s \mid N = n\} P\{N = n\} \\ &= \sum_{n=1}^{\infty} P\{\textstyle\sum_{i=1}^{n} Y_i = s\} P\{N = n\}. \end{aligned}$$

The expected value of S can be computed from

$$\begin{aligned} E(S) &= \sum_{n=1}^{\infty} E(S \mid N = n) P\{N = n\} \\ &= \sum_{n=1}^{\infty} E(\textstyle\sum_{i=1}^{n} Y_i) P\{N = n\} \\ &= \sum_{n=1}^{\infty} (\textstyle\sum_{i=1}^{n} E(Y_i)) P\{N = n\} \\ &= \mu E(N), \end{aligned}$$

where $\mu = E(Y)$.

1.3 DISCRETIZATION

Discrete stochastic structures are also useful for solving and analyzing continuous structures. In various problem setups, the continuous model is discretized to make the problem mathematically tractable. For example, in wind power analysis the continuos wind speed distribution is usually transformed into a discrete distribution, see, e.g. (Li and Zio 2012). Let V denote wind speed random variable and $f(v)$ be the probability density function of V. Assume that the wind speed distribution is discretized by defining s states. That is, we define a new discrete distribution that can take the values $v_1, v_2, ..., v_s$. The probability that the wind speed equals to v_i is computed from

$$p(v_i) = \begin{cases} \int\limits_{0}^{\frac{v_i+v_{i+1}}{2}} f(v)dv, & \text{if } i = 1 \\ \int\limits_{\frac{v_{i-1}+v_i}{2}}^{\frac{v_i+v_{i+1}}{2}} f(v)dv, & \text{if } i = 2, ..., s-1 \\ \int\limits_{\frac{v_{i-1}+v_i}{2}}^{\infty} f(v)dv, & \text{if } i = s. \end{cases}$$

In the case of a cubic relationship between the wind speed and the wind turbine output, we have the following wind turbine output when the wind speed is v:

$$g(v) = \begin{cases} 0, & \text{if } v < v_{ci} \text{ or } v \geq v_{co} \\ P_r \frac{(v^3 - v_{ci}^3)}{(v_r^3 - v_{ci}^3)}, & \text{if } v_{ci} \leq v < v_r \\ P_r, & \text{if } v_r \leq v < v_{co}, \end{cases}$$

where the wind turbine characteristics v_{ci}, v_r, and v_{co} respectively denote the cut-in wind speed, rated wind speed, and cut-out wind speed values (Louie and Sloughter 2014). P_r denotes the rated power of the wind turbine. For a wind farm that consists of n identical wind turbines, the wind farm output is

$$S_n \cdot g(V),$$

where V is the wind speed random variable and S_n denotes the total number of available wind turbines. If the availability/reliability of the wind turbine is denoted by p, then under the discretized wind speed model, the cumulative distribution function of wind farm output can be computed by conditioning on both S_n and V:

$$\begin{aligned} &P\{S_n \cdot g(V) \leq x\} \\ &= \sum_{l=0}^{n} \sum_{j=1}^{s} P\{l \cdot g(v_j) \leq x\} P\{V = v_j\} P\{S_n = l\} \\ &= \sum_{l=0}^{n} \sum_{j=1}^{s} u(\frac{x}{l} - g(v_j)) p(v_j) \binom{n}{l} p^l (1-p)^{n-l}, \end{aligned}$$

where $u(x) = 1$ if $x \geq 0$, and $u(x) = 0$ if $x < 0$ (Giorsetto and Utsurogi 1983). Similarly, the expected wind farm output can be computed from

$$E(S_n \cdot g(V)) = \sum_{l=0}^{n} \sum_{j=1}^{s} l g(v_j) p(v_j) \binom{n}{l} p^l (1-p)^{n-l}.$$

1.4 COMBINATORIAL METHODS

Combinatorial techniques have been effectively used in probability especially to derive the exact distribution of various discrete random variables. In particular, the exact distribution of a discrete random variable which appears as a function of integer-valued random sequence can be obtained via combinatorial arguments under certain assumptions. See, e.g. (Charalambides 2002). Consider the distribution given by (1.1) when $X_1, X_2, ..., X_n$ consists of independent and identical trials such that $p = P\{X_i = 1\}$ and $1 - p = P\{X_i = 0\}, i = 1, 2, ..., n$. The conditional distribution of R_n given $S_n = l$ can be written as

$$P\{R_n = r \mid S_n = l\} = \frac{N(n, r, l)}{\binom{n}{l}},$$

where $N(n, r, l)$ denotes the total number of binary sequences having l successes (ones) and r success runs. In fact, $N(n, r, l)$ denotes the number of binary sequences in the form of

$$\underbrace{0...0}_{y_1}\underbrace{1...1}_{z_1}\underbrace{0...0}_{y_2}\underbrace{1...1}_{z_2}...\underbrace{0...0}_{y_r}\underbrace{1...1}_{z_r}\underbrace{0...0}_{y_{r+1}},$$

where

$$y_1 + ... + y_{r+1} = n - l;\ y_1 \geq 0, y_2 > 0, ..., y_r > 0, y_{r+1} \geq 0, \tag{1.3}$$

and

$$z_1 + ... + z_r = l;\ z_1 > 0, ..., z_r > 0. \tag{1.4}$$

The number of integer solutions to (1.3) and (1.4) are given respectively by $\binom{n-l+1}{r}$ and $\binom{l-1}{r-1}$. Thus, we have

$$N(n, r, l) = \binom{n-l+1}{r}\binom{l-1}{r-1}.$$

Therefore,

$$P\{R_n = r\} = \sum_{l} \binom{n-l+1}{r}\binom{l-1}{r-1} p^l (1-p)^{n-l}.$$

For more details on runs and run related statistics, see, e.g. (Balakrishnan and Koutras 2001). ■

In the following example, the lifetime distribution of a discrete time cold standby repairable system is derived via combinatorial arguments. The detailed derivations were presented by Eryilmaz (2016).

Example 1.6. Consider a system consisting of one active and one cold standby component. The cold standby component has a zero failure rate and does not fail when it is inactive. At time $t = 0$, both of the two components are new, and component 1 starts work first, while component 2 is in a cold standby state. The standby component is put into operation upon the failure of the active component (component 1), and a repair action is taken for the component 1. The component is assumed to be as good as new after repair. Let $X_0, X_1, X_2, \ldots$ denote successive lifetimes of the active component, and R_i represents the repair time of a component upon the ith failure, $i = 1, 2, \ldots$. Then the system survives as soon as the lifetime of an active components exceeds the repair time of the previously failed component. That is, the lifetime of the system can be expressed as

$$T = \sum_{i=0}^{M} X_i,$$

where the random variable M is defined by

$$\{M = m\} \text{ iff } \{X_1 \geq R_1, \ldots, X_{m-1} \geq R_{m-1}, X_m < R_m\}.$$

Assume that

$$P\{X_i = x\} = p_1(1 - p_1)^{x-1} \text{ for } i = 0, 1, \ldots,$$

and

$$P\{R_i = x\} = p_2(1 - p_2)^{x-1} \text{ for } i = 1, 2, \ldots,$$

and $x = 1, 2, \ldots$, and the sequence of random variables $X_0, X_1, X_2, \ldots$ is independent of $R_1, R_2, \ldots$. Suppose that at the end of system's lifetime there are totally $i \leq n$ failures. The conditional event $\{T = n \mid R_1 = r_1, \ldots, R_{n-1} = r_{n-1}\}$ can be described as

$$\underbrace{0\ldots0}_{y_1 \geq 0}1 \; \underbrace{0\ldots0}_{y_2 \geq r_1 - 1} \; 1 \; \underbrace{0\ldots0}_{y_3 \geq r_2 - 1} \; 1\ldots1 \; \underbrace{0\ldots0}_{y_{i-1} \geq r_{i-2} - 1} \; 1 \; \underbrace{0..0}_{0 \leq y_i < r_{i-1} - 1} \; 1, \tag{1.5}$$

where $x_0 = y_1 + 1, x_1 = y_2 + 1, \ldots, x_{i-1} = y_i + 1$ represents successive lifetimes of active component. The probabilities of "1" and "0" in (1.5) are p_1 and $1 - p_1$, respectively. Then, by conditioning on i, Eryilmaz (2016) obtained the following equation:

$$\begin{aligned} &P\{T = n \mid R_1 = r_1, \ldots, R_{n-1} = r_{n-1}\} \\ &= \sum_{i=2}^{n} C(n, r_1, \ldots, r_{i-1}) p_1^i (1 - p_1)^{n-i}, \end{aligned}$$

for $n = 2, 3, \ldots$, where $C(n, r_1, \ldots, r_{i-1})$ is the number of binary sequences in the form of (1.5), and is given by

$$\begin{aligned} &C(n, r_1, \ldots, r_{i-1}) \\ &= \left[\binom{n - \sum_{j=1}^{i-2} r_j + i - 3}{i - 1} - \binom{n - \sum_{j=1}^{i-1} r_j + i - 2}{i - 1} \right]. \end{aligned}$$

By conditioning on $R_1,...,R_{n-1}$, the distribution of the system's lifetime is obtained as

$$P\{T=n\}=(n-1)p_1^2(1-p_1)^{n-2}+\sum_{i=0}^{n-2}p_1^{i+2}(1-p_1)^{n-i-2}$$
$$\times\left[\sum_{z=1}^{n-i-1}\binom{n-z-2}{i-1}\binom{z+i}{i+1}p_2^i(1-p_2)^{n-z-i-1}\right.$$
$$\left.-\sum_{z=1}^{n-i-1}\binom{n-z-1}{i}\binom{z+i}{i+1}p_2^{i+1}(1-p_2)^{n-z-i-1}\right], \qquad (1.6)$$

for $n=2,3,...$.

1.5 RECURSIVE EQUATIONS

The distribution given by (1.6) is exact and nonrecursive. That is, for given model parameters p_1 and p_2, the probabilities $P\{T=n\}$ can be directly computed for a given n. In some cases, it may not be possible to get a non-recursive equation for the distribution under concern. Indeed, in reliability engineering, recursive equations have been widely used for reliability calculations.

Example 1.7. (Eryilmaz 2012) considered a system which is subject to periodic random shocks such that a shock occurs with probability p at each period $n=1,2,...$. The magnitude of the shock which occurs at period j is a random variable denoted by B_j. The system fails if and only if the sum of the magnitudes of cumulative shocks exceeds, the level k for $k>0$. For $j\geq 1$, define

$$Y_j=\begin{cases} B_j, & \text{if a shock occurs at period } j \\ 0, & \text{otherwise}\end{cases}$$

Assuming $B_1,B_2,...$ to be a sequence of strictly positive random variables which are independent and identically distributed with probability mass function f_B, the failure time under the cumulative shock model is defined by the random variable

$$W_k=\min\left\{n:\sum_{j=1}^{n}Y_j>k\right\}.$$

(Eryilmaz 2012) obtained the following recursive equation for the distribution of W_k:

$$P\{W_k>n\}=p\sum_{b=1}^{\min(k,b^u)}P\{W_{k-b}>n-1\}f_B(b)+(1-p)P\{W_k>n-1\},$$

for $n\geq 1$ with $P\{W_k>0\}=1$, where b^u is the right hand limit of the support of f_B. The expected value of W_k has been found to be

$$E(W_k)=\frac{1}{p}+\sum_{b=1}^{\min(k,b^u)}E(W_{k-b})f_B(b), \qquad (1.7)$$

with $E(W_0) = \frac{1}{p}$. For an illustration, let

$$f_B(b) = (b-1)\alpha^2(1-\alpha)^{b-2}, \quad b = 2,3,...,$$

for $0 < \alpha < 1$. Then from (1.7),

$$E(W_k) = \frac{1}{p} + \alpha^2 \sum_{b=1}^{k} E(W_{k-b})(b-1)(1-\alpha)^{b-2}.$$

Assume that $p = 0.1$ and $\alpha = \frac{1}{2}$. Then,

$$E(W_5) = 20, \quad \text{and} \quad E(W_{10}) = 32.5.$$

1.6 SIMULATION

Many stochastic models that cannot be evaluated via analytical techniques may be analyzed by Monte Carlo simulation. The method of simulation can also be used to verify and check the accuracy of the analytical result. Monte Carlo simulation has been effectively used in solving engineering models. It is mainly based on generation of a random number from a probability distribution. Consider the problem of producing an outcome from Bernoulli experiment which has two possible outcomes as either "1" or "0". Let the probability of getting "1" be p, and the probability of getting "0" be $1-p$. If a random number that is generated from the uniform distribution on $(0,1)$ is denoted by u, then

$$\text{if } u \leq p \Longrightarrow x = 1,$$
$$\text{if } u > p \Longrightarrow x = 0,$$

where x denotes the outcome of the Bernoulli experiment.

Consider generating a random number from a given discrete distribution

$$p(x) = P\{X = x\}, \; x = x_1, x_2, ...,$$

where $\sum_{i=1}^{\infty} p(x_i) = 1$. Let u be a random number from the uniform distribution on $(0,1)$. Then,

$$x = \begin{cases} x_1, & \text{if } u < p_1 \\ x_2, & \text{if } p_1 < u < p_1 + p_2 \\ \vdots & \vdots \\ x_m, & \text{if } \sum_{i=1}^{m-1} p_i < u < \sum_{i=1}^{m} p_i \\ \vdots & \vdots \end{cases}$$

Example 1.8. Consider the problem of estimating the mean number of success runs (previously considered in section 1.4) by Monte Carlo simulation approach. For a sequence $X_1, X_2, ..., X_n$ that consists of independent and identical trials such that

$p = P\{X_i = 1\}$ and $1-p = P\{X_i = 0\}, i = 1,2,...,n$, total number of success runs in $X_1, X_2, ..., X_n$ can be defined as

$$R_n = \sum_{i=1}^{n} I(X_i = 1, X_{i+1} = 0), \tag{1.8}$$

where $P\{X_{n+1} = 0\} = 1$ and $I(A) = 1$ if A occurs and $I(A) = 0$, otherwise. To approximate $E(R_n)$, we first generate $\mathbf{X}^{(1)} = (X_1^{(1)}, X_2^{(1)}, ..., X_n^{(1)})$ and then compute the value of (1.8) based on $\mathbf{X}^{(1)}$. Let $R_n^{(1)}$ denote the value of (1.8) for $\mathbf{X}^{(1)}$. Similarly, a random vector $\mathbf{X}^{(2)} = (X_1^{(2)}, X_2^{(2)}, ..., X_n^{(2)})$ is generated and $R_n^{(2)}$ is computed based on $\mathbf{X}^{(2)}$. The same procedure is repeated m times and $R_n^{(1)}, R_n^{(2)}, ..., R_n^{(m)}$ are computed. By the strong law of large numbers, we have

$$\lim_{m \longrightarrow \infty} \frac{R_n^{(1)} + R_n^{(2)} + ... + R_n^{(m)}}{m} = E(R_n^{(i)}).$$

Thus, we can use the mean of generated sequence $R_n^{(1)}, R_n^{(2)}, ..., R_n^{(m)}$ as an estimate of $E(R_n)$. That is, for sufficiently large m,

$$E(R_n) \simeq \frac{1}{m} \sum_{i=1}^{m} R_n^{(i)}.$$

From (1.7), the exact value of $E(R_n)$ can be computed from

$$\begin{aligned} E(R_n) &= (n-1)P\{X_i = 1, X_{i+1} = 0\} + P\{X_i = 1\} \\ &= (n-1)p(1-p) + p. \end{aligned}$$

Example 1.9. The problem is to estimate the survival probability and mean time to failure of the parallel system with two independent components. Let X_i denote the lifetime of the ith component such that $P\{X_i = t\} = p(t)$, $i = 1,2$. Clearly, the parallel system fails when both components fail and hence the lifetime of the system is $T = \max(X_1, X_2)$. For a given t, to approximate $P\{T > t\}$, we first generate $\mathbf{X}^{(1)} = (X_1^{(1)}, X_2^{(1)})$ and then compute $T^{(1)} = \max(X_1^{(1)}, X_2^{(1)})$. Note that $X_1^{(1)}$ and $X_2^{(1)}$ are independently generated from the distribution with probability mass function $p(t)$. Similarly, $\mathbf{X}^{(2)} = (X_1^{(2)}, X_2^{(2)})$ is generated and $T^{(2)} = \max(X_1^{(2)}, X_2^{(2)})$ is computed based on $\mathbf{X}^{(2)}$. The same procedure is repeated m times and $T^{(1)}, T^{(2)}, ..., T^{(m)}$ are computed. For sufficiently large m, the survival function and mean time to failure of the parallel system can be approximated respectively by

$$P\{T > t\} \simeq \frac{1}{m} \sum_{i=1}^{m} I\left(T^{(i)} > t\right),$$

and

$$E(T) \simeq \frac{1}{m} \sum_{i=1}^{m} T^{(i)},$$

where $I(A) = 1$ if A occurs and $I(A) = 0$ otherwise. For an illustration, let

$$p(t) = p(1-p)^{t-1},\ t = 1,2,...,$$

and $p = 0.1$. Then, from the simulation based on $m = 10{,}000$ trials we obtain

$$\begin{aligned} P\{T > 2\} &\simeq 0.9633, \\ P\{T > 3\} &\simeq 0.9281, \\ E(T) &\simeq 14.6908. \end{aligned}$$

An analytical solution for this reliability problem will be developed in section 6.

Example 1.10. A linear consecutive-k-out-of-n:F system is a system that consists of n linearly ordered components and fails if at least k consecutive components fail. Assume that the components are independent and the reliability of the ith component is p, $i = 1,2,...,n$. We define a binary random variable X_i that represents the state of the ith component such that

$$X_i = \begin{cases} 1, & \text{if the } i\text{th component is working} \\ 0, & \text{if the } i\text{th component has failed.} \end{cases}$$

Let L_n be the length of the longest failure run in $X_1, X_2, ..., X_n$. Then, the reliability of the linear consecutive-k-out-of-n:F system can be represented as

$$R(p) = P\{L_n < k\}. \tag{1.9}$$

The problem is to compute the reliability $R(p)$ for given p, n, and k. This problem has been extensively studied in reliability literature and various methods have been developed for calculating (1.9). The details are presented in sections 2 and 8. To approximate (1.9) by the method of Monte Carlo simulation, the following algorithm can be used. We first generate a binary sequence $\mathbf{X}^{(1)} = (X_1^{(1)}, X_2^{(1)}, ..., X_n^{(1)})$ and then compute the value of the longest failure run $L_n^{(1)}$ based on $\mathbf{X}^{(1)}$. A random vector of binary sequence $\mathbf{X}^{(2)} = (X_1^{(2)}, X_2^{(2)}, ..., X_n^{(2)})$ is generated and $L_n^{(2)}$ is computed based on $\mathbf{X}^{(2)}$. The same procedure is repeated m times and $L_n^{(1)}, L_n^{(2)}, ..., L_n^{(m)}$ are obtained. For sufficiently large m, the reliability of the system can be approximated by

$$R(p) \simeq \frac{1}{m}\sum_{i=1}^{m} I\left(L_n^{(i)} < k\right).$$

The length of the longest failure run of a binary sequence $X_1, X_2, ..., X_n$ can be obtained in the following way. Let's define

$$Y_i = \begin{cases} 0, & \text{if } X_i = 1 \\ Y_{i-1} + 1, & \text{if } X_i = 0, \end{cases}$$

for $i = 1,2,...,n$, where $Y_0 = 0$. Then,

$$L_n = \max(Y_1, ..., Y_n).$$

For an illustration, we approximate $R(p)$ for selected values of p, n, and k:

$$\begin{aligned}
n = 20,\ k = 2 &\Longrightarrow R(0.95) = 0.9549,\ R(0.97) = 0.9849,\\
n = 20,\ k = 3 &\Longrightarrow R(0.95) = 0.9977,\ R(0.97) = 0.9994,\\
n = 30,\ k = 3 &\Longrightarrow R(0.95) = 0.9971,\ R(0.97) = 0.9992.
\end{aligned}$$

Example 1.11. Let X and Y be two independent random variables such that

$$P\{X = x\} = p(1-p)^{x-1},\ x = 1, 2, \ldots,$$

for $0 < p < 1$, and

$$P\{Y = y\} = \theta(1-\theta)^{y-1},\ y = 1, 2, \ldots,$$

for $0 < \theta < 1$. Consider the probability $P\{X < Y\}$. This probability can be analytically computed as

$$\begin{aligned}
P\{X < Y\} &= \sum_{y=1}^{\infty} P\{X < y\} P\{Y = y\}\\
&= \sum_{y=1}^{\infty} \left[1 - (1-p)^{y-1}\right] \theta(1-\theta)^{y-1}\\
&= 1 - \frac{\theta}{1-(1-p)(1-\theta)}.
\end{aligned}$$

To find an approximation via simulation, the following method can be used. We successively generate $(X^{(1)}, Y^{(1)}), \ldots, (X^{(m)}, Y^{(m)})$. Then,

$$P\{X < Y\} \simeq \frac{1}{m} \sum_{i=1}^{m} I\left(X^{(i)} < Y^{(i)}\right).$$

The values of $P\{X < Y\}$ are given below for selected values of the parameters p and θ:

$$\begin{aligned}
p = 0.5, \theta = 0.5 &\Longrightarrow P\{X < Y\} = 0.3333,\\
p = 0.5, \theta = 0.1 &\Longrightarrow P\{X < Y\} = 0.8182,\\
p = 0.25, \theta = 0.1 &\Longrightarrow P\{X < Y\} = 0.6923.
\end{aligned}$$

REFERENCES

1. An, Z-W., Huang, H-Z., Liu, Y. 2008. A discrete stress–strength interference model based on universal generating function. *Reliability Engineering and System Safety* 93: 1485-1490.
2. Balakrishnan, N. and Koutras, M.V. 2001. *Runs and scans with applications*. Wiley Series in Probability and Statistics, New York.
3. Charalambides, C.A. 2002. *Enumerative Combinatorics*, Chapman & Hall/CRC.

4. Choi, D.I., Knessl, C. and Tier, C. A. 1999. Queueing system with queue length dependent service times with applications to cell discarding in ATM networks. *Journal of Applied Mathematics and Stochastic Analysis* 12: 35-62.
5. Eryilmaz, S. 2012. Life behavior of a system under discrete shock model. *Discrete Dynamics in Nature and Society*, 238641.
6. Eryilmaz, S. 2015. Assessment of a multi-state system under a shock model. *Applied Mathematics and Computation* 269: 1-8.
7. Eryilmaz, S. 2016. Discrete time cold standby repairable system: Combinatorial analysis. *Communications in Statistics-Theory and Methods* 45: 7399-7405.
8. Giorsetto, P. and Utsurogi, K.F. 1983. Development of a new procedure for reliability modeling of wind turbine generators. *IEEE Transactions on Power Apparatus and Systems* PAS-102: 134-143.
9. Krishna, H. and Pundir, P.S. 2009. A bivariate geometric distribution with applications to reliability. *Communications in Statistics-Theory and Methods* 38: 1079-1093.
10. Li, Y.F and Zio, E. 2012. A multi-state model for the reliability assessment of a distributed generation system via universal generating function. *Reliability Engineering and System Safety* 106: 28-36.
11. Louie, H. and Sloughter, J.M. 2014. Probabilistic modeling and statistical characteristics of aggregate wind power. *Large Scale Renewable Power Generation, Green Energy and Technology*, 19-51.
12. Oliveira, R.P., Achcar, J.A., Peralta, D. and Mazucheli, J. 2019. Discrete and continuous bivariate lifetime models in presence of cure rate: a comparative study under Bayesian approach. *Journal of Applied Statistics* 46: 449-467.
13. Phatak, A. G. and Sreehari, M. 1981. Some characterizations of a bivariate geometric distribution. *J. Ind. Statist. Assoc.* 19: 141-146.

2 Discrete Probability Distributions

Most of the discrete probability distributions are based on a sequence of Bernoulli trials. A Bernoulli sequence $X_1, X_2, \ldots$ consists of independent and identical trials such that

$$P\{X_i = 1\} = p \text{ and } P\{X_i = 0\} = 1 - p,\ i = 1, 2, \ldots$$

That is, the Bernoulli trials appear as outcomes of a random experiment with two possible outcomes as either a "success" (denoted by "1") or a "failure" (denoted by "0"). The terms "success" and "failure" are generic and can be replaced by appropriate concepts depending on the experiment/model/system/process under concern. For example, in a stochastic system that consists of n components/units, the random variable X_i may denote the state of the ith component, where

$$X_i = \begin{cases} 1, & \text{if the } i\text{th component is operating} \\ 0, & \text{if the } i\text{th component has failed,} \end{cases}$$

$i = 1, 2, \ldots, n$. In this case, $p = P\{X_i = 1\}$ denotes the working probability (reliability) of the ith component. In statistical quality control, a typical process consists of computing a statistic that is based on samples/subgroups and comparing it with control limits. Let LCL and UCL denote respectively the fixed lower control limit and upper control limit. Let's define a Bernoulli trial such that

$$X_i = \begin{cases} 1, & \text{if } Y_i \notin (LCL, UCL) \\ 0, & \text{if } Y_i \in (LCL, UCL), \end{cases}$$

where Y_i is a statistic that is based on the ith subgroup. Then, $p = P\{Y_i \notin (LCL, UCL)\}$ represents the probability of having a point outside the control limits.

In a certain reliability demonstration test, successive tests are applied to an equipment. Assume that the outcome of each individual test can be either success or failure. Then,

$$X_i = \begin{cases} 1, & \text{if the } i\text{th test is successful} \\ 0, & \text{if the } i\text{th test is failure} \end{cases}$$

In this case, $p = P\{X_i = 1\}$ denotes the probability that the ith test is successful.

2.1 BINOMIAL MODEL

The Bernoulli trial corresponds to a specific event. That is, for an event A, we can define $X_i = 1$ if A occurs and $X_i = 0$ if A^c occurs, where A^c denotes the complement of the event A. Counting the number of occurrences of a specific event is of special

DOI: 10.1201/9781003160632-2

importance in many problem setups. For a sequence of Bernoulli trials $X_1, X_2, \ldots$, let S_n denote the number of successes (or equivalently the number of occurrences of the event A) in n trials. That is, $S_n = \sum_{i=1}^{n} X_i$. Because the trials $X_1, X_2, \ldots$ are independent and have the same probability of success $p = P\{X_i = 1\}, i = 1, 2, \ldots$, the probability of having s successes in n trials is obtained as

$$P\{S_n = s\} = \binom{n}{s} p^s (1-p)^{n-s}, \ s = 0, 1, \ldots, n. \tag{2.1}$$

The random variable S_n whose probability mass function is given by (2.1) is said to have a Binomial distribution. The mean number of successes (or the mean number of occurrences of the event A) can be computed from

$$E(S_n) = E\left(\sum_{i=1}^{n} X_i\right) = \sum_{i=1}^{n} E(X_i) = np,$$

where $p = E(X_i) = P\{X_i = 1\}, i = 1, 2, \ldots, n$. The variance of S_n is

$$Var(S_n) = Var\left(\sum_{i=1}^{n} X_i\right) = np(1-p).$$

Example 2.1. Consider a wind plant that consists of n identical wind turbines. Let p denote the availability of each turbine. The availability is in fact the probability of proper operation of the wind turbine. The term availability is used instead of reliability since a wind turbine typically consists of repairable units. In power system reliability analysis, $1-p$ is known to be forced outage rate (FOR). Thus, at any instant of time, the number of available wind turbines can be modeled by a Binomial distribution. For example, the probability that exactly s wind turbines are available in the wind plant can be computed from (2.1). For a wind plant consisting of $n = 10$ wind turbines, assume that the FOR for each turbine is $1-p = 0.04$. Then, at any instant of time, the probability that all wind turbines are available is

$$P\{S_{10} = 10\} = p^{10} = (0.96)^{10} = 0.6648.$$

The probability that at least 8 wind turbines are available can be computed from

$$P\{S_{10} \geq 8\} = \sum_{s=8}^{10} \binom{10}{s} (0.96)^s (0.04)^{10-s} = 0.9938. \quad \blacksquare$$

(Klenke and Mattner 2010) presented stochastic ordering results on classical discrete distributions with respect to their parameters. For two random variables X and Y, X is said to be stochastically smaller than Y (denoted by $X \leq_{st} Y$) if $P\{X \geq x\} \leq P\{Y \geq x\}$ for all x. Let $S_n(p)$ be a Binomial random variable with parameters n and p. Then, among other stochastic ordering results, (Klenke and Mattner 2010) proved the following:

$$S_{n_1}(p_1) \leq_{st} S_{n_2}(p_2) \text{ if and only if } (1-p_1)^{n_1} \geq (1-p_2)^{n_2} \text{ and } n_1 \leq n_2.$$

Example 2.2. Suppose that the system A consists of $n_1 = 3$ independent components of type A such that each component of type A works with probability $p_1 = 0.96$ during a fixed period of time. System B consists of $n_2 = 4$ independent components of type B and each component of type B works with probability $p_2 = 0.93$. Assume that each system works properly if at least two of its components work. Let $S_{n_1}(p_1)$ $(S_{n_2}(p_2))$ denote the number of working components within the system A (B). Because $n_1 \leq n_2$ and $6.4 \times 10^{-5} = (1-p_1)^{n_1} \geq (1-p_2)^{n_2} = 2.401 \times 10^{-5}$, we have $S_{n_1}(p_1) \leq_{st} S_{n_2}(p_2)$ which implies $P\{S_{n_1}(p_1) \geq 2\} \leq P\{S_{n_2}(p_2) \geq 2\}$. Therefore, the system B has a larger working probability than the system A. Indeed,

$$P\{S_{n_1}(p_1) \geq 2\} = \sum_{i=2}^{3} \binom{3}{i} (0.96)^i (0.04)^{3-i} = 0.9953,$$

and

$$P\{S_{n_2}(p_2) \geq 2\} = \sum_{i=2}^{4} \binom{4}{i} (0.93)^i (0.07)^{4-i} = 0.9987. \quad \blacksquare$$

Example 2.3. The failed machines arrive at the maintenance center for repair at time marks. Assume that no more than one machine may arrive at a given time mark and the arrivals occur independently. The probability of arrival of a machine at a time mark is p and the probability of no arrival at a time mark is $1-p$. Then, the number of machines arriving within a given period of time follows Binomial distribution. In particular, for a time period t_k containing k time marks, the probability of m arrivals within t_k is

$$\binom{k}{m} p^k (1-p)^{k-m}, \ m = 0, 1, ..., k. \quad \blacksquare$$

Consider the following modification of the Binomial model. Suppose that the trials are classified in two groups as type I and type II. Let $n = n_1 + n_2$, and n_i denote the number of trials of type i, $i = 1, 2$. The trials are no longer identical but are assumed to be independent. Denote by p_i the probability of success for the trials of type i, $i = 1, 2$. If $S_{n_1}^{(1)}$ and $S_{n_2}^{(2)}$ denote respectively the total number of successes of type I and type II, then

$$\begin{aligned} P&\left\{S_{n_1}^{(1)} = m_1, S_{n_2}^{(2)} = m_2\right\} \\ &= \binom{n_1}{m_1}\binom{n_2}{m_2} p_1^{m_1}(1-p_1)^{n_1-m_1} p_2^{m_2}(1-p_2)^{n_2-m_2}, \end{aligned}$$

for $m_1 = 0, 1, ..., n_1$; $m_2 = 0, 1, ..., n_2$. In this case, the probability of having s successes in $n = n_1 + n_2$ trials is

$$\begin{aligned} P\{S_n = s\} &= \sum_m P\left\{S_{n_1}^{(1)} = s - m, S_{n_2}^{(2)} = m\right\} \\ &= \sum_{m=\max(0, s-n_1)}^{\min(s, n_2)} \binom{n_1}{s-m}\binom{n_2}{m} p_1^{s-m}(1-p_1)^{n_1-s+m} p_2^m (1-p_2)^{n_2-m}, \end{aligned}$$

for $s = 0, 1, ..., m$. Manifestly, the mean number of successes in $n = n_1 + n_2$ trials is

$$E(S_n) = n_1 p_1 + n_2 p_2.$$

Example 2.4. Consider a wind plant consisting of $n_1 = 4$ wind turbines of type I and $n_2 = 6$ wind turbines of type II. Suppose that the FOR values for the turbines of type I and type II are given respectively by $1 - p_1 = 0.04$ and $1 - p_2 = 0.03$. Then, at any instant of time, the probability that at least 8 wind turbines are available can be computed from

$$\begin{aligned} P\{S_{10} \geq 8\} &= \sum_{s=8}^{10} \sum_{m=\max(0,s-4)}^{\min(s,6)} \binom{4}{s-m} \binom{6}{m} (0.96)^{s-m} (0.04)^{4-s+m} \\ &\times (0.97)^m (0.03)^{6-m} \\ &= 0.9960. \quad \blacksquare \end{aligned}$$

Next, consider the more general case when n trials are nonidentical. Let p_i denote the probability of success for the ith trial, i.e. $p_i = P\{X_i = 1\}, i = 1, 2, ..., n$. We aim to find the distribution of $S_n = \sum_{i=1}^n X_i$ for independent but nonidentical Bernoulli trials $X_1, X_2, ...$ In the case of independent but nonidentical Bernoulli trials, the distribution of S_n is known to be Poisson's Binomial distribution (Wang 1993).

(Wang 1993) obtained the following equation for the probability mass function of the Poisson's Binomial distribution:

$$P\{S_n = s\} = \sum_{A \in \Im_s} \left(\prod_{i \in A} p_i \right) \left(\prod_{j \in A^c} (1 - p_j) \right),$$

where $A \in \Im_s$ implies that $A = (i_1, ..., i_s)$ is an ordered set such that $i_j < i_k$ if $j < k$.(Wang 1993) has also obtained the following alternative expression:

$$P\{S_n = s\} = \prod_{j=1}^{n} (1 - p_j) \left[\sum_{A \in \wp_s} \left(\prod_{i \in A} \left(\frac{p_i}{1 - p_i} \right) \right) \right] / s!,$$

where

$$\wp_s = \{\pi(A) : A \in \Im_s, \pi \text{ is a permutation in } \Re^s\}.$$

Under the nonidentical setup, we have

$$E(S_n) = \sum_{i=1}^{n} p_i.$$

For given n and s, the probability $P\{S_n = s\}$ can be computed by using a recursive equation. Indeed, by using the total probability law, for $n \geq s$,

$$\begin{aligned} P\{S_n = s\} &= P\{S_n = s, X_n = 1\} + P\{S_n = s, X_n = 0\} \\ &= P\{S_{n-1} = s-1\} P\{X_n = 1\} + P\{S_{n-1} = s\} P\{X_n = 0\} \\ &= P\{S_{n-1} = s-1\} p_n + P\{S_{n-1} = s\} (1 - p_n), \end{aligned}$$

and $P\{S_n = s\} = 0$ if $n < s$ with $P\{S_0 = 0\} = 1, P\{S_1 = 1\} = p_1$. For an illustration, let $n = 3$ and $s = 2$. Then,

$$P\{S_3 = 2\} = P\{S_2 = 1\}\, p_3 + P\{S_2 = 2\}\,(1 - p_3),$$

$$\begin{aligned} P\{S_2 = 2\} &= P\{S_1 = 1\}\, p_2 + P\{S_1 = 2\}\,(1 - p_2) \\ &= p_1 p_2, \end{aligned}$$

$$\begin{aligned} P\{S_2 = 1\} &= P\{S_1 = 0\}\, p_2 + P\{S_1 = 1\}\,(1 - p_2) \\ &= (1 - p_1)p_2 + p_1(1 - p_2) \\ &= p_1 + p_2 - 2p_1 p_2. \end{aligned}$$

Thus, we have

$$\begin{aligned} P\{S_3 = 2\} &= p_3(p_1 + p_2 - 2p_1 p_2) + p_1 p_2(1 - p_3) \\ &= p_1 p_2 + p_1 p_3 + p_2 p_3 - 3p_1 p_2 p_3. \quad \blacksquare \end{aligned}$$

(Charalambides 2010) studied q-distributions on Bernoulli trials with a geometrically varying success probability. For a sequence $X_1, \ldots, X_n$ of binary trials with possible outcomes "0" and "1", the trials of the subsequence after the $(i-1)$st "0" until the ith "0" are assumed to be independent with failure probability

$$q_i = 1 - \theta q^{i-1},\ i = 1, 2, \ldots,\ 0 < \theta < 1, 0 < q \leq 1.$$

Under this model, the distribution of the random variable $S_n = \sum_{i=1}^{n} X_i$ is known to be q-binomial distribution. (Charalambides 2010) obtained the following expression for the probability mass function of the q-binomial distribution:

$$P\{S_n = s\} = \begin{bmatrix} n \\ s \end{bmatrix}_q \theta^s \prod_{i=1}^{n-s} (1 - \theta q^{i-1}),$$

for $s = 0, 1, \ldots, n, 0 < q < 1$, where

$$\begin{bmatrix} n \\ s \end{bmatrix}_q = \frac{[n]_{s,q}}{[s]_q!},$$

and

$$\begin{aligned} [x]_{k,q} &= [x]_q [x-1]_q \cdots [x-k+1]_q, \\ [x]_q &= (1 - q^x)/(1 - q), \\ [x]_q! &= [1]_q [2]_q \cdots [x]_q. \end{aligned}$$

Manifestly,

$$\begin{bmatrix} n \\ s \end{bmatrix}_q \to \binom{n}{s}$$

for $q \to 1$. Thus, the q-binomial distribution converges to the usual binomial distribution as $q \to 1$.

2.2 WAITING TIME MODELS

The well-known geometric distribution is the simplest of the waiting time distributions. It is the distribution of the random variable that counts the number of trials until the first success in a sequence of Bernoulli trials. If the corresponding random variable is denoted by W, then the sequence that satisfies $\{W = x\}$ has the form $\underbrace{00...0}_{x-1}1$.

For a sequence of Bernoulli trials $X_1, X_2, ...,$ the random variable W may be formally defined as

$$W = \min(n : X_n = 1).$$

Thus, the probability mass function of W is

$$P\{W = x\} = p(1-p)^{x-1}, x = 1, 2, ... \tag{2.2}$$

For a geometric random variable that has probability mass function (2.2), it can be easily seen that

$$P\{W > t + s \mid W > t\} = P\{W > s\}, \tag{2.3}$$

for all $t, s > 0$. The equation (2.3) implies the "memoryless" property of the geometric distribution. As an immediate consequence of (2.3), we have

$$E(W - t \mid W > t) = E(W) = \frac{1}{p}.$$

If the random variable W is defined based on a sequence of nonidentical Bernoulli trials, then

$$P\{W = x\} = p_x \prod_{i=1}^{x-1} (1 - p_i), \; x = 2, 3, ...,$$

and $P\{W = 1\} = p_1$.

Example 2.5. In statistical process control, the number of samples/subgroups that needs to be collected before the first out-of-control signal is given by a control chart, which is called a run-length random variable. A charting statistic is computed for each collected sample/subgroup. Suppose that the process is declared to be out-of-control when a single point (charting statistic) is beyond the fixed control limits LCL and UCL. If the successive charting statistics $Y_1, Y_2, ...$ are obtained independently and they have the same sampling distribution, then the run length random variable has a geometric distribution with mean

$$\frac{1}{P\{Y_i \notin (LCL, UCL)\}},$$

where Y_i is the charting statistic that is based on the ith sample. ■

Example 2.6. Suppose that a system is subject to random shocks over time. A shock can be seen as a stress such as thermal, mechanical, and electrical. According to the extreme shock model, the system breaks down when the magnitude of an individual shock exceeds some given level c. That is, the system failure occurs due to an extreme large shock. Let $Y_1, Y_2, ...$ denote the magnitudes of successive shocks (or the damages due to shocks). If $Y_1, Y_2, ...$ are independent and identically distributed, then the number W of shocks until the failure of the system has geometric distribution with probability mass function

$$P\{W = x\} = P\{Y_i > c\} [P\{Y_i \leq c\}]^{x-1} \ , x = 1, 2, \dots .$$

In this case, the mean number of shocks that cause failure of the system is

$$E(W) = \frac{1}{P\{Y_i > c\}}. \quad \blacksquare$$

Example 2.7. For the service system example described in Example 2.3, the interarrival times between successive arrivals follow geometric distribution with mean $\frac{1}{p}$. ■

Example 2.8. Consider a system consisting of two serially connected independent components. Let W_i denote the lifetime (time to failure) of the ith component, $i = 1, 2$ and $P\{W_i = t\} = p(1-p)^{t-1}$. Manifestly, the lifetime of the series system is $\min(W_1, W_2)$. Denote by N the number of failed components at the time when the system fails. Then,

$$\begin{aligned} P\{N = 1\} &= P\{W_1 < W_2\} + P\{W_2 < W_1\} \\ &= 2P\{W_1 < W_2\} \\ &= 2\sum_{t=1}^{\infty} P\{W_1 = t\} P\{W_2 > t\} \\ &= 2\sum_{t=1}^{\infty} p(1-p)^{t-1}(1-p)^t \\ &= 2\frac{(1-p)}{2-p}, \end{aligned}$$

and

$$\begin{aligned} P\{N = 2\} &= P\{W_1 = W_2\} \\ &= \sum_{t=1}^{\infty} P\{W_1 = t\} P\{W_2 = t\} \\ &= \sum_{t=1}^{\infty} \left[p(1-p)^{t-1}\right]^2 \\ &= \frac{p}{2-p}. \end{aligned}$$

Thus, the mean number of failed components at the time when the system fails is found to be

$$E(N) = \frac{2}{2-p}.$$

Clearly, $E(N) \to 1$ as $p \to 0$ and $E(N) \to 2$ as $p \to 1$. ■

The probability generating function is a useful tool to study distributional properties corresponding to waiting time models. For a discrete random variable W, it is defined by

$$\phi(z) = E(z^W) = \sum_x z^x P\{W = x\}$$

for $0 < z < 1$. For the geometric distribution given by (2.2),

$$\phi(z) = \frac{pz}{1-(1-p)z}.$$

The expected value of random variable W can be computed from

$$E(W) = \phi'(1).$$

A more general waiting time model is based on the number of trials until r $(r \geq 1)$ successes in a sequence of Bernoulli trials. Let W_r denote the total number of trials until the occurrence of r successes in $X_1, X_2, \ldots$. A typical sequence for the occurrence of the event $\{W_r = x\}$ ends with the success at the xth trial, and the preceding $x-1$ trials include exactly $r-1$ successes. Thus,

$$P\{W_r = x\} = \binom{x-1}{r-1} p^r (1-p)^{x-r}, \; x = r, r+1, \ldots \tag{2.4}$$

A random variable whose probability mass function is given by (2.4) is said to have a negative binomial distribution. A geometric distribution is a special case of a negative binomial distribution with $r = 1$. If $Z_1, Z_2, \ldots$ are independent geometric random variables with common probability mass function (2.2), then

$$W_r = \sum_{i=1}^{r} Z_i. \tag{2.5}$$

That is, a negative binomial random variable can be represented as a sum of independent geometric random variables. By using representation (2.5), the probability generating function and the expected value of W_r are obtained respectively as

$$E(z^{W_r}) = \left[\frac{pz}{1-(1-p)z}\right]^r,$$

and

$$E(W_r) = \frac{r}{p}.$$

Example 2.9. Assume that the number of flights before the replacement of tires of jet fighters is a random variable that has negative binomial distribution having the probability mass function (2.4) with $r = 2$ and $p = \frac{1}{15}$. Then, the mean number of flights until the replacement of tires is $E(W_2) = 30$. The probability that the tires are replaced after the tenth flight is

$$P\{W_2 > 10\} = 1 - \sum_{x=2}^{10} (x-1)p^2(1-p)^{x-2} = 0.8599. \quad \blacksquare$$

In negative binomial distribution, the stopping time depends on the specified number r of successes. The clustering of the successes is irrelevant and only the total number of successes is recorded. In some problem setups, the occurrence of a specified number of consecutive successes might be more adequate. That is, a success run of a fixed length may cause stopping. For example, a system that is subject to external shocks may break down due to k consecutive critical shocks. In statistical quality control, the process is declared to be out-of-control when the values of k consecutive charting statistics fall beyond the control limits.

Let T_k denote the total number of trials until k consecutive successes in a sequence of Bernoulli trials $X_1, X_2, \ldots$ with $p = P\{X_i = 1\}, i = 1, 2, \ldots$. For an illustration, consider the Bernoulli sequence 10011010111. Manifestly, $T_1 = 1, T_2 = 5$, and $T_3 = 11$. To make clear the definition of T_k, we compute probabilities $P\{T_k = x\}$ for selected values of x when $k = 2$ by considering all possible binary sequences satisfying $\{T_2 = x\}$:

$$P\{T_2 = 2\} = P\{11\} = p^2,$$

$$P\{T_2 = 3\} = P\{011\} = p^2(1-p),$$

$$P\{T_2 = 4\} = P\{0011\} + P\{1011\} = p^2(1-p)^2 + p^3(1-p),$$

$$\begin{aligned} P\{T_2 = 5\} &= P\{00011\} + P\{10011\} + P\{01011\}, \\ &= p^2(1-p)^3 + 2p^3(1-p)^2. \end{aligned}$$

The random variable T_k may be formally defined as

$$T_k = \min\{n : X_{n-k+1} = \ldots = X_n = 1\}.$$

Clearly,

$$P\{T_k = x\} = \begin{cases} 0, & \text{if } x < k \\ p^k, & \text{if } x = k \\ (1-p)p^k, & \text{if } x = k+1. \end{cases} \tag{2.6}$$

On the other hand, for $x > k+1$, the event $\{T_k = x\}$ occurs if and only if the last $k+1$ trials are $0\underbrace{1\ldots1}_{k}$, and no k consecutive successes occur in the first $x-k-1$ trials, i.e. $\{T_k > x-k-1\}$. Because the trials are independent,

$$\begin{aligned} P\{T_k = x\} &= (1-p)p^k P\{T_k > x-k-1\} \\ &= (1-p)p^k \left[1 - \sum_{i=k}^{x-k-1} P\{T_k = i\}\right], \end{aligned} \tag{2.7}$$

for $x > k+1$. It immediately follows from (2.7) that

$$P\{T_k = x-1\} - P\{T_k = x\} = (1-p)p^k P\{T_k = x-k-1\},$$

or equivalently

$$P\{T_k = x\} = P\{T_k = x-1\} - (1-p)p^k P\{T_k = x-k-1\}, \tag{2.8}$$

for $x > k+1$. Thus, the probability mass function of T_k is computed recursively using (2.6) and (2.8).

The distribution of the random variable T_k is known to be geometric distribution of order k in the literature. For $k = 1$, we have geometric distribution. The geometric distribution of order k has been of great interest in numerous practical applications, ranging from quality control and reliability to non-parametric statistics.

Alternative expressions for $P\{T_k = x\}$ have been obtained by applying different methods. For example, (Muselli 1996) derived the following combinatorial expression:

$$P\{T_k = x\}$$
$$= \sum_{j=1}^{\left[\frac{x+1}{k+1}\right]} (-1)^{j-1} p^{jk} (1-p)^{j-1} \left[\binom{x-jk-1}{j-2} + (1-p)\binom{x-jk-1}{j-1} \right], \tag{2.9}$$

for $x > k$, where $[a]$ denotes the integer part of a. Matrix-based mathematically more efficient formula for the probability mass function of T_k will be presented in Chapter 4.

The probability generating function of T_k is given by

$$E(z^{T_k}) = \frac{(pz)^k (1-pz)}{1 - z + (1-p) p^k z^{k+1}}, \tag{2.10}$$

see, e.g. (Philippou et al. 1983), (Balakrishnan and Koutras 2001). The expected value of T_k is

$$E(T_k) = \frac{1-p^k}{(1-p)p^k} \tag{2.11}$$

which can be obtained by differentiating (2.10) and evaluating it at $z = 1$.

Example 2.10. A start-up demonstration test is used to elicit information about the reliability of a unit before purchasing it. Successive attempts to start (start-up) the unit are made and the outcomes are observed as either a successful start-up or failure. The decision about the acceptance/rejection of the unit is made based on certain rules. According to the simplest rule, the unit is accepted when k consecutive successful start-ups is obtained (Hahn and Gage 1983). In this case, the test length is in fact the number of start-ups until k consecutive successful start-ups. Thus, the test length random variable has geometric distribution of order k.

The parameter p denotes the probability of having a successful start-up. For an illustration, if $k = 5$ and $p = 0.9$, then using (2.11), the mean test length, i.e. the mean number of start-ups until the acceptance of the unit is found to be

$$E(T_5) = \frac{1-(0.9)^5}{0.1(0.9)^5} \simeq 7$$

(when the resulting value is rounded to the nearest integer).

Example 2.11. Consider an oil transportation system for transporting oil from the point A to the point B. There are n equally spaced pump stations between the points

A and B. Suppose that each pump station is able to transport oil to the next k stations. Thus, the oil transportation is interrupted if at least k consecutive pump stations fail, i.e. cannot pump the oil. This is a system that consists of n linearly ordered components (pump stations). The system fails if there are at least k consecutively failed components. Such a system is known to be linear consecutive-k-out-of-n:F system in the literature. Let "1" and "0" represent respectively the failure and operational states for the components. For $n = 5$ and $k = 2$, if the states of the components are 10101, then the system is in a working state. For the same values of n and k, if the states of the components are 10011, then the system is in a failure state. If p denotes the common failure probability of pump stations, then the reliability of the system can be calculated from $P\{T_k > n\}$, where T_k denotes the number of trials until k consecutive ones in $X_1, X_2, ...$ with $X_i = 1$ if the ith component fails and $X_i = 0$ if the ith component works. Manifestly, the event $\{T_k > n\}$ implies that no k or more consecutive failures occur in $X_1, ..., X_n$. Let the pump stations be independent and $n = 30, k = 3, p = 0.03$, then the system works properly with probability

$$P\{T_3 > 30\} = 1 - \sum_{x=3}^{30} P\{T_3 = x\} = 0.9993. \quad \blacksquare$$

Various modifications and generalizations of geometric distribution of order k have been defined and studied in the literature. A more general waiting time random variable is the number of trials until the rth occurrence of k consecutive successes. The resulting distribution is called negative Binomial distribution of order k [(Charalambides 1986), (Philippou and Georghiou 1989), (Antzoulakos and Philippou 1999)]. Under nonoverlapping scheme, the distribution of the number of trials until the rth appearance of k consecutive successes is the r-fold convolution of geometric distribution of order k when the outcomes are classical Bernoulli trials. Thus, if the corresponding waiting time random variable is denoted by $T_{r,k}$, then its probability generating function is

$$E(z^{T_{r,k}}) = \left[\frac{(pz)^k(1-pz)}{1-z+(1-p)p^k z^{k+1}}\right]^r.$$

The mean of $T_{r,k}$ is

$$E(T_{r,k}) = r\frac{1-p^k}{(1-p)p^k}.$$

(Pervez et al. 2017) presented an application of negative Binomial distribution of order k in sensitive surveys. The distribution of $T_{r,k}$ has been studied by (Koutras 1997) and (Antzoulakos and Chadjiconstantinidis 2001) for binary sequences consisting of Markov dependent trials.

(Eryilmaz 2014) introduced the geometric distribution of order k with a reward. Let $X_1, X_2, ...$ be a sequence of independent trials with two possible outcomes as either a success "1" or failure "0" with $P\{X_i = 1\} = p_i$. Suppose that each time a success occurs a reward is received. Let Y_i denote the reward earned at the time of the ith success, where $P\{Y_i > 0\} = 1$. Although the random variable Y_i is called as a reward, it can have a different meaning such as amount, magnitude, loss or gain depending on the model. Suppose that $Y_1, Y_2, \ldots$ is a sequence of independent random variables with $F_i(x) = P\{Y_i \leq x\}, i = 1, 2, \ldots$. The sequence $Y_1, Y_2, \ldots$ is also assumed to be independent of $X_1, X_2, \ldots$. Let's define T_k^R as the number of trials until the sum of consecutive Y_is is equal to or exceeds the level k, $k > 0$. The distribution of the waiting time random variable T_k^R has been called geometric distribution of order k with a reward.

Clearly, the distribution of T_k^R is same with the geometric distribution of order k when $P\{Y_i = 1\} = 1$ for all i. Let

$$S_{b-a+1}^{(a.b)} = \Sigma_{i=a}^{b} Y_i \ , \ \bar{F}_{b-a+1}^{(a.b)}(k) = P\left\{S_{b-a+1}^{(a.b)} \geq k\right\},$$

for $1 \leq a \leq b$. Then, for $x > 1$,

$$P\{T_k^R = x\} = \sum_{n=1}^{x-1} (\bar{F}_n^{(x-n+1.x)}(k) - \bar{F}_{n-1}^{(x-n+1.x-1)}(k))(1 - p_{x-n})$$

$$\times \prod_{i=x-n+1}^{x} p_i \left[1 - \sum_{i=0}^{x-n-1} P\{T_k^R = i\}\right] + (\bar{F}_x^{(1.x)}(k) - \bar{F}_{x-1}^{(1.x-1)}(k))\prod_{i=1}^{x} p_i$$

and $P\{T_k^R = x\} = p_1 \bar{F}_1^{(1.1)}(k)$ for $x = 1$ (Eryilmaz 2014).

Example 2.12. Let p_i be the probability of occurrence of heavy rainfall on the day i in a specific location, $i = 1, 2, \ldots$. A soft rainfall is absorbed by the system while the certain number of consecutive days of extremely heavy rain may result flooding in the location. In this setup, Y_i denotes the flood volume caused by the heavy precipitation on the day i. Thus, under this setup, the random variable T_k^R represents the number of days until the sum of the flood volumes of consecutive days is equal to or exceeds the level k for $k > 0$. ■

2.3 POISSON MODEL

The Poisson distribution may be obtained as a limiting distribution of the binomial distribution and can be used for modeling the number of occurrences of an event in continuous time. Indeed, if the number of trials n in binomial distribution is very large, i.e. $n \to \infty$ and the probability of success p is very small, i.e. $p \to 0$, then

$$\binom{n}{s} p^s (1-p)^{n-s} \to \frac{e^{-\lambda}\lambda^s}{s!},$$

for $s = 0, 1, \ldots$, where

$$\lambda = \lim_{n\to\infty} np.$$

The random variable ξ whose probability mass function is given by (2.12) is said to have a Poisson distribution.

$$P\{\xi = s\} = \frac{e^{-\lambda}\lambda^s}{s!}, \ s = 0, 1, \ldots \tag{2.12}$$

The expected value of ξ is $E(\xi) = \lambda$.

Example 2.13. Consider a power system consisting of a specified number of transformers. The mean number of transformer failures is known to be 30 for a given year. If ξ denotes the total number of failures in a given year, then the probability that exactly 35 failures occur in a year is given by

$$P\{\xi = 35\} = \frac{e^{-30}30^{35}}{35!} = 0.0453. \quad \blacksquare$$

In the following, we consider the determination of the optimal quantity that will be produced by the supplier in the context of discrete inventory optimization. Let ξ denote the demand for a specified period of time. Assume that the random variable ξ follows Poisson distribution with probability mass function (2.12). Let's define Q as the quantity that should be produced and

b: Sales price per unit

c_o: the (unit) overage cost (the cost incurred as a result of ordering too much inventories)

c_u: the (unit) underage cost (the profit lost as a result of not having enough inventory)

c: the (unit) cost of the product.

Then, for given demand $\xi = d$, the profit is (Vandeput 2020):

$$\wp(Q,d) = b\min(Q,d) - c_o\max(0,Q-d) - c_u\max(0,d-Q) - cQ.$$

Thus, the expected profit is

$$\begin{aligned} E[\wp(Q,\xi)] &= \sum_d \wp(Q,d)P\{\xi = d\} \\ &= b\sum_d \min(Q,d)P\{\xi = d\} - c_o\sum_d \max(0,Q-d)P\{\xi = d\} \\ &\quad - c_u\sum_d \max(0,d-Q)P\{\xi = d\} - cQ. \end{aligned}$$

Clearly,

$$\begin{aligned} \sum_d \min(Q,d)P\{\xi = d\} &= \sum_{d \le Q} d \cdot P\{\xi = d\} + QP\{\xi > Q\} \\ &= \sum_{d=0}^{Q} d \cdot P\{\xi = d\} + Q\left[1 - \sum_{d=0}^{Q} P\{\xi = d\}\right] \\ &= \sum_{d=0}^{Q} d \cdot \frac{e^{-\lambda}\lambda^d}{d!} + Q\left[1 - \sum_{d=0}^{Q} \frac{e^{-\lambda}\lambda^d}{d!}\right], \end{aligned}$$

$$\begin{aligned} \sum_d \max(0,Q-d)P\{\xi = d\} &= \sum_{d<Q} (Q-d)P\{\xi = d\} \\ &= \sum_{d=0}^{Q-1} (Q-d)\frac{e^{-\lambda}\lambda^d}{d!}, \end{aligned}$$

and

$$\sum_d \max(0, d-Q)P\{\xi = d\} = \sum_{d>Q}(d-Q)P\{\xi = d\}$$
$$= \sum_{d=Q+1}^{\infty}(d-Q)\frac{e^{-\lambda}\lambda^d}{d!}.$$

Thus, the expected profit becomes

$$E[\wp(Q,\xi)] = b\sum_{d=0}^{Q} d\cdot\frac{e^{-\lambda}\lambda^d}{d!} + bQ\left[1-\sum_{d=0}^{Q}\frac{e^{-\lambda}\lambda^d}{d!}\right]$$
$$-c_o\sum_{d=0}^{Q-1}(Q-d)\frac{e^{-\lambda}\lambda^d}{d!} - c_u\sum_{d=Q+1}^{\infty}(d-Q)\frac{e^{-\lambda}\lambda^d}{d!} - cQ.$$

The problem is to find Q^* that maximizes $E[\wp(Q,\xi)]$ for given b, c_o, c_u, c, and λ.

One of the most important stochastic processes for modeling the number of occurrences of a randomly occurring event over continuous time is the Poisson process. Consider occurrences of a random event A over time. Let $N(t)$ represent the number of the occurrences of the event A by time t. The counting process $N(t)$ is called a Poisson process with rate $\lambda > 0$, if (i) $N(0) = 0$, (ii) the process $N(t)$ has independent increments, i.e. the numbers of events that occur in disjoint time intervals are independent, and (iii) the number of events in any interval of length t is Poisson distributed with mean λt. Thus, the one dimensional distribution of the process $N(t)$ is given by

$$P\{N(t) = n\} = \frac{e^{-\lambda t}(\lambda t)^n}{n!}, \quad n = 0, 1, \ldots$$

for all $t \geq 0$. Clearly, $E(N(t)) = \lambda t$.

If S_n denotes the arrival time of the nth event in the process, then the following dual relationship holds true:

$$P\{N(t) \geq n\} = P\{S_n \leq t\}.$$

The Poisson process has been widely used in reliability engineering. See, e.g. (Cha and Finkelstein 2018), (Zhao and Nakagawa 2018). In the following, the use of Poisson process is illustrated with a spare part inventory problem.

The problem is to find the optimal number of spare parts inventories based on a certain criteria. Such a problem has been studied in the literature from different perspectives. (Louit et al. 2011) presented optimization models for critical spare parts inventories via reliability approach. The methodology that is based on the Poisson process proposed by (Louit et al. 2011) is summarized in the following. Consider a specific component used by a certain equipment or a system. A group of m independent components is required to operate for an interval of length T. That is, m equipments are used independently. Let μ be the mean time to replacement for the component under concern. Let's define $N(T,m)$ as the total number of replacements in the interval $[0,T]$ and $S(k,m)$ be the time until the kth replacement. Then,

$$P\{N(T,m) < k\} = P\{S(k,m) > T\}.$$

If the demand for a spare part follows a Poisson process with replacement rate λ, then $\lambda = \frac{1}{\mu}$. That is,

$$\lambda = \frac{1}{MTBF},$$

where $MTBF$ denotes the mean time between failures. Assume that the failed component is non-repairable. We aim to find the stock level required to ensure that no stock-outs occur over a selected interval of time. The demand process corresponding to m independent Poisson processes is in fact the superposed process. Thus, the probability of having s demands in $[0,T]$ can be calculated from

$$P\{N(T,m) = s\} = \frac{e^{-a}a^s}{s!},$$

where a is the expected number of demands in $[0,T]$, i.e.

$$a = m\lambda T = \frac{mT}{\mu}.$$

If r denotes the probability that a spare part will not be rejected due to lack of spares in stock, then for a given reliability value r, the optimal stock level k^* can be obtained such that

$$P\{N(T,m) \leq k^*\} = P\{S(k^*+1,m) > T\} = \sum_{s=0}^{k^*} \frac{e^{-a}a^s}{s!} \geq r.$$

2.4 COMPOUND MODELS

For a discrete random variable N with probability mass function $P\{N = n\}$ and a sequence of independent and identically distributed random variables $Y_1, Y_2, ...,$ the compound random variable defined by $S = \sum_{i=1}^{N} Y_i$ has attracted substantial research interest. If the sequence of random variables $Y_1, Y_2, ...$ has common discrete distribution and independent of N, then the probability generating function of S is

$$\begin{aligned} E(z^S) &= \sum_n E(z^{\sum_{i=1}^n Y_i})P\{N=n\} \\ &= \sum_n \left[E(z^Y)\right]^n P\{N=n\} \\ &= \phi_N(\psi(z)), \end{aligned} \tag{2.13}$$

where $\phi_N(z) = E(z^N)$ and $\psi(z) = E(z^Y)$ denote respectively the probability generating functions of N and Ys. By differentiating (2.13) and evaluating it at $z = 1$, the expected value of S is found to be

$$E(S) = E(N)E(Y). \tag{2.14}$$

The distribution of the random variable S is usually named with respect to the distribution of N. One of the most frequently considered cases is the case when N has Poisson distribution. In this case, the distribution of S is called compound Poisson. If N has Poisson distribution with mean λ, then

$$\phi_N(z) = E(z^N) = e^{-\lambda(1-z)}.$$

Thus, the probability generating function of S is

$$E(z^S) = e^{-\lambda(1-\psi(z))}.$$

Example 2.14. Consider a system that is subject to shocks over time. A shock that occurs at a random time can be either fatal or non-fatal. Let $Y_1, Y_2, ...$ be the times between arrivals of shocks. The system breaks down upon the occurrence of a fatal shock. Let p_1 denote the probability of a fatal shock. Assume that the interarrival times $Y_1, Y_2, ...$ between successive shocks follow geometric distribution with mean $\frac{1}{p_2}$. Then, the system's lifetime is $S = \sum_{i=1}^{N} Y_i$, where

$$P\{N = n\} = p_1(1-p_1)^{n-1},$$

and

$$P\{Y_i = y\} = p_2(1-p_2)^{y-1}, i = 1, 2, ...$$

In this case, the random variable N represents the number of shocks until the first fatal shock. Using (2.13), the probability generating function of the system's lifetime is obtained as

$$E(z^S) = \frac{p_1 p_2 z}{1-(1-p_1 p_2)z}$$

which implies that the system's lifetime has geometric distribution with mean $\frac{1}{p_1 p_2}$. That is,

$$P\{S = s\} = p_1 p_2 (1-p_1 p_2)^{s-1}, \ s = 1, 2, ...$$

The expected value of S can be alternatively obtained from (2.14) as

$$E(S) = E(N)E(Y) = \frac{1}{p_1 p_2}. \quad \blacksquare$$

(Koutras and Eryilmaz 2017) studied the distribution of S by choosing the distribution of N as geometric distribution of order k. The resulting distribution is called compound geometric distribution of order k. For notational convenience, let $N \stackrel{st}{=} T_k$ and the random variable T_k has probability generating function (2.10). Then, the probability generating function of S is

$$E(z^S) = \frac{(p\psi(z))^k(1-p\psi(z))}{1-\psi(z)+(1-p)p^k(\psi(z))^{k+1}},$$

where $\psi(z)$ is the probability generating function of discrete random variable Y.

Example 2.15. For a system that is subject to shocks over time, let $Y_1, Y_2, \ldots$ be the times between arrivals of shocks. The magnitude of the ith shock is denoted by Z_i, $i \geq 1$. The system fails if the magnitudes of k consecutive shocks exceed or equal to a prespecified level $c > 0$. Such a shock model is known to be run shock model in the literature (see, e.g. Mallor and Omey 2001). The lifetime of the system is then defined by the random variable $S = \sum_{i=1}^{T_k} Y_i$, where T_k has geometric distribution of order k with $p = P\{Z \geq c\}$. Assume that the system fails upon the occurrence of $k = 2$ consecutive shocks whose magnitudes are at least c. The probability p is known to be $p = 0.1$. If the mean time between successive shocks is $E(Y) = 120$ hours, then the mean time to failure of the system is

$$E(S) = E(T_2)E(Y) = \frac{1+p}{p^2}E(Y) = 13{,}200 \text{ hours.} \quad \blacksquare$$

2.5 MODELS VIA DISCRETIZATION OF CONTINUOUS DISTRIBUTIONS

Discrete analogues of various well-known continuous distributions have been defined and studied in the literature. For a positive continuous random variable ξ, define $\eta = [\xi] + 1$, where $[a]$ is the integer part of a. Then, the probability mass function of η is

$$\begin{aligned} P\{\eta = m\} &= P\{m-1 \leq \xi < m\} \\ &= F_\xi(m) - F_\xi(m-1), \end{aligned} \tag{2.15}$$

where $F_\xi(m)$ is the cumulative distribution function of ξ.

Example 2.16. Let ξ have exponential distribution with cumulative distribution function $F_\xi(m) = 1 - e^{-\lambda m}$. Then, from (2.15)

$$P\{\eta = m\} = e^{-\lambda(m-1)}(1 - e^{-\lambda}),$$

for $m = 1, 2, \ldots$ which is the probability mass function of geometric random variable with parameter $p = 1 - e^{-\lambda}$.

Example 2.17. Let ξ have Weibull distribution with $F_\xi(m) = 1 - e^{-\left(\frac{m}{\beta}\right)^\alpha}; \alpha, \beta > 0$. Then,

$$P\{\eta = m\} = q^{(m-1)^\alpha} - q^{m^\alpha}, \tag{2.16}$$

for $m = 1, 2, \ldots$, where $q = e^{-\frac{1}{\beta^\alpha}}$. The distribution given by (2.16) has been named as discrete Weibull distribution and extensively studied by (Nakagawa and Osaki 1975).

Example 2.18. Consider a system which is subject to a sequence of shocks that arrive according to a Polya process. That is, if $N(t)$ denotes the number of shocks arrived in $(0,t)$, then

$$P\{N(t) = n\} = \binom{\alpha+n-1}{n}\left(\frac{t}{t+\beta}\right)^n\left(\frac{\beta}{\beta+t}\right)^\alpha, \tag{2.17}$$

for $n = 0, 1, ...$ Polya process is a mixed Poisson process, and hence (2.17) can be written as

$$P\{N(t) = n\} = \int_0^\infty \frac{e^{-\lambda t}(\lambda t)^n}{n!} dH(\lambda),$$

where

$$dH(\lambda) = \frac{\beta^\alpha}{\Gamma(\alpha)} \lambda^{\alpha-1} e^{-\beta\lambda} d\lambda$$

(Teugels and Vynckier 1996). The system fails when the time between two consecutive shocks falls below a fixed threshold δ. This is known to be δ-shock model in the literature (see, e.g. Li and Kong 2007). Because the shocks occur according to a Polya process, the interarrival times $Y_1, Y_2, ...$ between successive shocks are conditionally independent and

$$P\{Y_1 > y, ...Y_m > y\} = \int_0^\infty e^{-m\lambda y} dH(\lambda) = \left(\frac{\beta}{my+\beta}\right)^\alpha.$$

Let η be the number of shocks until the failure of the system. Then,

$$\begin{aligned} P\{\eta = m\} &= P\{Y_1 > \delta, ..., Y_{m-1} > \delta, Y_m \leq \delta\} \\ &= P\{Y_1 > \delta, ..., Y_{m-1} > \delta\} - P\{Y_1 > \delta, ..., Y_m > \delta\} \\ &= \left(\frac{1}{1+(m-1)\frac{\delta}{\beta}}\right)^\alpha - \left(\frac{1}{1+m\frac{\delta}{\beta}}\right)^\alpha, \end{aligned} \tag{2.18}$$

for $m = 1, 2, ...$ (Eryilmaz 2017). The distribution given by (2.18) is discrete Pareto distribution and can be obtained from continuous Pareto distribution as $\eta = [\xi] + 1$, where $F_\xi(m) = 1 - \left(\frac{q}{q+m}\right)^\alpha$, where $q = \frac{\beta}{\delta}$. ■

For more examples on discrete analogues of well-known continuous distributions, see, e.g. (Kemp 1997), (Roy 2004), (Kemp 2004), (Chakraborty and Chakravarty 2012), (Bakouch 2014), (Abouammoh and Alhazzani 2015).

2.6 BIVARIATE BINOMIAL AND GEOMETRIC MODELS

Consider a sequence of independent bivariate random vectors $(X_1^{(1)}, X_1^{(2)}), (X_2^{(1)}, X_2^{(2)})$, ... such that

$$p_{00} = P\left\{X_i^{(1)} = 0, X_i^{(2)} = 0\right\},$$

$$p_{10} = P\left\{X_i^{(1)} = 1, X_i^{(2)} = 0\right\},$$

$$p_{01} = P\left\{X_i^{(1)} = 0, X_i^{(2)} = 1\right\},$$

$$p_{11} = P\left\{X_i^{(1)} = 1, X_i^{(2)} = 1\right\},$$

for all $i \geq 1$. We define

$$S_n^{(1)} = \sum_{i=1}^{n} X_i^{(1)}, \; S_n^{(2)} = \sum_{i=1}^{n} X_i^{(2)}.$$

That is, $S_n^{(j)}$ denotes the total number of ones (successes) in the first n trials of $X_1^{(j)}, X_2^{(j)}, \ldots,\ j = 1,2$. Then, $(S_n^{(1)}, S_n^{(2)})$ has a bivariate Binomial distribution with the joint probability mass function

$$\begin{aligned} &P\left\{S_n^{(1)} = k_1, S_n^{(2)} = k_2\right\} \\ &= \sum_{a=\max(0,n-k_1-k_2)}^{\min(n-k_1,n-k_2)} \frac{n!}{a!(n-a-k_2)!(n-a-k_1)!(k_1+k_2+a-n)!} p_{00}^{a} p_{10}^{n-a-k_2} \\ &\times p_{01}^{n-a-k_1} p_{11}^{k_1+k_2+a-n}, \end{aligned} \tag{2.19}$$

for $n \geq \max(k_1, k_2)$ (Johnson, Kotz and Balakrishnan 1997). For more details on bivariate Binomial model and its modifications, see, e.g. (Kocherlakota and Kocherlakota 1992), (Bayramoglu and Kemalbay 2013).

As $n \to \infty$, the limiting distribution of the bivariate Binomial distribution given by (2.19) when

$$p_{10} = \frac{\lambda_{10}}{n}, \; p_{01} = \frac{\lambda_{01}}{n}, \; p_{11} = \frac{\lambda_{11}}{n}$$

is the bivariate Poisson distribution whose probability mass function is (Kawamura 1973),

$$P\{X = k_1, Y = k_2\} = \sum_{\delta=0}^{\min(k_1,k_2)} \frac{\lambda_{10}^{k-\delta} \lambda_{01}^{l-\delta} \lambda_{11}^{\delta}}{(k-\delta)!(l-\delta)!\delta!} e^{-(\lambda_{10}+\lambda_{01}+\lambda_{11})}.$$

Let W_j denote the waiting time for the first success in $X_1^{(j)}, X_2^{(j)}, \ldots,\ j = 1,2$. Then (W_1, W_2) has a bivariate geometric distribution with the joint probability mass function (Marshall and Olkin 1985):

$$\begin{aligned} &P\{W_1 = t_1, W_2 = t_2\} \\ &= \begin{cases} p_{00}^{t_1-1} p_{10}(1-p_{01}-p_{11})^{t_2-t_1-1}(p_{01}+p_{11}), & \text{if} \quad t_1 < t_2 \\ p_{11} p_{00}^{t_1-1}, & \text{if} \quad t_1 = t_2 \\ p_{00}^{t_2-1} p_{01}(1-p_{10}-p_{11})^{t_1-t_2-1}(p_{10}+p_{11}), & \text{if} \quad t_1 > t_2. \end{cases} \end{aligned} \tag{2.20}$$

Example 2.19. Suppose that a system with two components is subject to a sequence of shocks. Assume that neither of the components are affected by the shock with probability p_{00}, the first component is affected but the second is not with probability p_{10}, the second component is affected but the first is not with probability p_{01}, and both components are affected by the shock with probability p_{11}. Let W_j represent the number of shocks until failure of the component $j, j = 1,2$. If the components fail as soon as they are affected by a single shock, then (W_1, W_2) has a bivariate geometric distribution with the probability mass function (2.20). The probability that the first component fails before the second component is

$$\begin{aligned}
P\{W_1 < W_2\} &= \sum\sum_{t_1<t_2} P\{W_1 = t_1, W_2 = t_2\} \\
&= \sum_{t_1=1}^{\infty} \sum_{t_2=t_1+1}^{\infty} p_{00}^{t_1-1} p_{10}(1-p_{01}-p_{11})^{t_2-t_1-1}(p_{01}+p_{11}) \\
&= \frac{p_{10}}{1-p_{00}}.
\end{aligned}$$

Example 2.20. Consider the shock model previously defined in Example 1.2. It can be easily seen that the joint distribution of (Z, W) is bivariate geometric and is same with (2.20) when

$$p_{00} = p_1 \ , \ p_{10} = p_2 \ , \ p_{01} = p_3 \ , \ p_{11} = 0.$$

Example 2.21. Assume that the joint behavior of the lifetimes of two dependent components is modeled by (2.20) with $p_{11} = 0$. That is,

$$\begin{aligned}
&P\{W_1 = t_1, W_2 = t_2\} \\
&= \begin{cases} p_{00}^{t_1-1} p_{10}(1-p_{01})^{t_2-t_1-1} p_{01}, & \text{if} \quad t_1 < t_2 \\ p_{00}^{t_2-1} p_{01}(1-p_{10})^{t_1-t_2-1} p_{10}, & \text{if} \quad t_1 > t_2. \end{cases}
\end{aligned}$$

If the components have a parallel connection, then the lifetime of the entire system is

$$T = \max(W_1, W_2).$$

The distribution of the system's lifetime can be computed as follows.

$$\begin{aligned}
P\{T \le t\} &= P\{W_1 \le t, W_2 \le t\} \\
&= \sum_{t_1=1}^{t-1} \sum_{t_2=t_1+1}^{t} P\{W_1 = t_1, W_2 = t_2\} \\
&\quad + \sum_{t_2=1}^{t-1} \sum_{t_1=t_2+1}^{t} P\{W_1 = t_1, W_2 = t_2\}.
\end{aligned}$$

Consider the two possible cases when $t_1 < t_2$ and when $t_1 > t_2$:

$$\begin{aligned}
P\{T \le t\} &= P\{W_1 \le t, W_2 \le t\} \\
&= \sum_{t_1=1}^{t-1} \sum_{t_2=t_1+1}^{t} p_{00}^{t_1-1} p_{10}(1-p_{01})^{t_2-t_1-1} p_{01} \\
&\quad + \sum_{t_2=1}^{t-1} \sum_{t_1=t_2+1}^{t} p_{00}^{t_2-1} p_{01}(1-p_{10})^{t_1-t_2-1} p_{10}.
\end{aligned}$$

We have

$$\sum_{t_1=1}^{t-1}\sum_{t_2=t_1+1}^{t} p_{00}^{t_1-1} p_{10}(1-p_{01})^{t_2-t_1-1} p_{01}$$
$$= p_{10}\left[\frac{(1-p_{01})^t}{p_{00}+p_{01}-1} + \frac{1-p_{00}^{t-1}}{1-p_{00}} + \frac{p_{00}^{t-1}(p_{01}-1)}{p_{00}+p_{01}-1}\right],$$

and

$$\sum_{t_2=1}^{t-1}\sum_{t_1=t_2+1}^{t} p_{00}^{t_2-1} p_{01}(1-p_{10})^{t_1-t_2-1} p_{10}$$
$$= p_{01}\left[\frac{(1-p_{10})^t}{p_{00}+p_{10}-1} + \frac{1-p_{00}^{t-1}}{1-p_{00}} + \frac{p_{00}^{t-1}(p_{10}-1)}{p_{00}+p_{10}-1}\right].$$

Thus, the distribution of T is obtained as

$$P\{T \le t\} = p_{10}\left[\frac{(1-p_{01})^t}{p_{00}+p_{01}-1} + \frac{1-p_{00}^{t-1}}{1-p_{00}} + \frac{p_{00}^{t-1}(p_{01}-1)}{p_{00}+p_{01}-1}\right]$$
$$+ p_{01}\left[\frac{(1-p_{10})^t}{p_{00}+p_{10}-1} + \frac{1-p_{00}^{t-1}}{1-p_{00}} + \frac{p_{00}^{t-1}(p_{10}-1)}{p_{00}+p_{10}-1}\right],$$

for $t > 1$, and $P\{T \le 1\} = 0$.

2.7 DISCRETE RELIABILITY INDICES

For a nonnegative random variable T with probability mass function $p(t) = P\{T = t\}$, the failure rate (hazard rate) is defined as

$$r(t) = \frac{P\{T = t\}}{P\{T \ge t\}} = \frac{p(t)}{\sum_{k \ge t} p(k)}.$$

For the geometric distribution with $p(t) = p(1-p)^{t-1}, t = 1, 2, \ldots$, we have

$$r(t) = \frac{p(1-p)^{t-1}}{(1-p)^{t-1}} = p.$$

That is, the hazard rate is constant and this result is parallel with the memoryless property of the geometric distribution.

The monotonicity of hazard rate function is of special importance in modeling failure time data. (Gupta et al. 1997) established the following results for the monotonicity of the hazard rate. Let's define

$$\eta(t) = \frac{p(t) - p(t+1)}{p(t)},$$

and

$$\Delta\eta(t) = \eta(t+1) - \eta(t).$$

Then,

(i) If $\Delta\eta(t) > 0$, then $r(t)$ is nondecreasing, Increasing failure rate (IFR).
(ii) If $\Delta\eta(t) < 0$, then $r(t)$ is nonincreasing, Decreasing failure rate (DFR).
(iii) If $\Delta\eta(t) = 0$ for all t, then $r(t)$ is constant.

Example 2.22. Assume that the number of flights for the failure of the jet tire follows the distribution with probability mass function:

$$p(t) = tp^2(1-p)^{t-1}, t = 1, 2, \ldots$$

Then,

$$r(t) = \frac{tp^2}{tp+1-p}.$$

Clearly,

$$\eta(t) = 1 - (1-p)\frac{t+1}{t},$$

and

$$\Delta\eta(t) = \eta(t+1) - \eta(t) = (1-p)\frac{1}{t(t+1)} > 0$$

which implies that $r(t)$ is IFR. It is also clear that $r(\infty) = p$.

Example 2.23. Consider the discrete Weibull distribution defined in Example 2.17. The hazard rate of the distribution is obtained as

$$r(t) = 1 - q^{t^\alpha - (t-1)^\alpha},$$

for $t = 1, 2, \ldots$. The function $r(t)$ is nondecreasing (IFR) if $\alpha \geq 1$, nonincreasing (DFR) if $0 < \alpha \leq 1$, and constant if $\alpha = 1$. ■

Another useful reliability measure is mean residual life function which is defined as

$$m(t) = E(T - t \mid T \geq t).$$

We have the following relation between the hazard rate and the mean residual life function:

$$r(t) = 1 - \frac{m(t)}{1 + m(t+1)}.$$

(Gupta 2015) proved that if T has IFR, then T has decreasing mean residual life function.

Example 2.24. For the model in Example 2.22, the mean residual life function is obtained as

$$m(t) = \frac{(1-p)(2-p+pt)}{p(1-p+pt)}.$$

Clearly,

$$m(t) > m(t+1) \text{ for all } t$$

which implies that T has decreasing mean residual life. For $p = 0.05$, given that the tire has survived for ten flights, the mean residual lifetime of the tire is computed as

$$\begin{aligned} m(10) &= E(T - 10 \mid T \geq 10) \\ &= \frac{(1-0.05)(2-0.05+0.05\cdot 10)}{0.05(1-0.05+0.05\cdot 10)} \\ &\simeq 32. \end{aligned}$$

2.8 DISCRETE RENEWAL PROCESS

Consider a component whose lifetime is measured in discrete units. Let T denote the lifetime of the component, and

$$P\{T = t\} = p_t,\ t = 1,2,..., \tag{2.21}$$

and

$$F(t) = P\{T \leq t\}.$$

Assume that the failed component is replaced by a new fresh one. We define $M(n)$ as expected number of replacements up to time n. That is, if $N(n)$ denotes the total number of replacements up to time n, then

$$M(n) = E(N(n)). \tag{2.22}$$

Suppose that the kth replacement occurs at time S_k. Then, we have

$$M(n) = \sum_{k=1}^{\infty} P\{S_k \leq n\}, \tag{2.23}$$

where

$$S_k = \sum_{i=1}^{k} T_i,$$

and $T_1, T_2, ...$ is a sequence of independent and identically distributed random variables with common distribution given by (2.21). The random variable T_i is in fact the lifetime of the ith component. The function defined by (2.22) is known to be a renewal function. Considering the representation (2.23), the renewal function can be computed as follows:

$$\begin{aligned} M(n) &= \sum_{k=1}^{\infty} P\{S_k \leq n\} \\ &= P\{T_1 \leq n\} + \sum_{k=2}^{\infty} P\left\{T_1 + \sum_{i=2}^{k} T_i \leq n\right\}. \end{aligned}$$

By conditioning on T_1,

$$\begin{aligned} M(n) &= F(n) + \sum_{k=2}^{\infty}\sum_{t=1}^{n-1} P\left\{\sum_{i=2}^{k} T_i \le n-t\right\} P\{T_1 = t\} \\ &= F(n) + \sum_{t=1}^{n-1}\sum_{k=2}^{\infty} P\left\{\sum_{i=2}^{k} T_i \le n-t\right\} P\{T_1 = t\} \\ &= F(n) + \sum_{t=1}^{n-1}\sum_{k=2}^{\infty} P\{S_{k-1} \le n-t\} P\{T_1 = t\} \\ &= F(n) + \sum_{t=1}^{n-1}\sum_{k=1}^{\infty} P\{S_k \le n-t\} P\{T_1 = t\} \\ &= F(n) + \sum_{t=1}^{n-1} M(n-t) p_t. \end{aligned}$$

Thus, for given lifetime distribution $p_t = P\{T = t\}$, the renewal function is computed from the following equation:

$$M(n) = F(n) + \sum_{t=1}^{n-1} M(n-t) p_t. \tag{2.24}$$

Discrete time renewal process has been used for modeling the arrival pattern of the demands in certain systems. See, e.g. (Lian et al. 2005), (Samanta et al. 2009).

Example 2.25. Assume that the times between replacements follow the distribution with probability mass function

$$p_t = p(1-p)^{t-1},\ t = 1, 2, \ldots$$

Then, from (2.24) we have

$$M(n) = 1 - (1-p)^n + \sum_{t=1}^{n-1} M(n-t) p(1-p)^{t-1}$$

which gives

$$M(n) = np.$$

REFERENCES

1. Antzoulakos, D. L. and Chadjiconstantinidis, S. 2001. Distributions of numbers of success runs of fixed length in Markov dependent trials. *Annals of the Institute of Statistical Mathematics* 53: 599-619.
2. Antzoulakos, D. L. and Philippou, A. N. 1999. Multivariate Pascal polynomials of order k with probability applications. In Applications of Fibonacci Numbers (Eds., G. E. Bergum et al.), 8: 27-41, Kluwer Academic Publishers, Amsterdam, The Netherlands.
3. Abouammoh, A.M. and Alhazzani, N.S. 2015. On discrete Gamma distribution. *Communications in Statistics-Theory and Methods* 44: 3087-3098.

4. Bakouch, H.S., Jazi, M.A. and Nadarajah, S. 2014. A new discrete distribution. *Statistics* 48: 200-240.
5. Balakrishnan, N. and Koutras, M.V. 2001. *Runs and scans with applications*. Wiley Series in Probability and Statistics, New York.
6. Bayramoglu, I. and Kemalbay, G. 2013. Some novel discrete distributions under fourfold sampling schemes and conditional bivariate order statistics. *Journal of Computational and Applied Mathematics* 248: 1-14.
7. Charalambides, Ch. A. 1986. On discrete distributions of order k. *Annals of the Institute of Statistical Mathematics* 38: 557-568.
8. Cha, J.H. and Finkelstein, M. 2018. *Point processes for reliability analysis*. Springer Series in Reliability Engineering.
9. Chakraborty, S. and Chakravarty, D. 2012. Discrete Gamma distributions: Properties and parameter estimations. *Communications in Statistics-Theory and Methods* 41: 3301-3324.
10. Charalambides, C.A. 2010. The q-Bernstein basis as a q-binomial distributions. *Journal of Statistical Planning and Inference* 140: 2184-2190.
11. Eryilmaz, S. 2014. Geometric distribution of order k with a reward. *Statistics and Probability Letters* 92: 53-58.
12. Eryilmaz, S. 2017. δ-shock model based on Polya process and its optimal replacement policy. *European Journal of Operational Research* 263: 690-697.
13. Gupta, P.L., Gupta, R.C. and Tripathi, R.C. 1997. On the monotonic properties of discrete failure rates. *Journal of Statistical Planning and Inference* 65: 255-268.
14. Gupta, P.L. 2015. Properties of reliability functions of discrete distributions. *Communications in Statistics-Theory and Methods* 44: 4114-4131.
15. Hahn, G. J. and Gage, J. B. 1983. Evaluation of a start-up demonstration test. *Journal of Quality Technology* 15: 103-106.
16. Johnson, N. L., Kotz, S. and Balakrishnan, N. 1997. *Discrete multivariate distributions*. New York: John Wiley & Sons.
17. Kawamura, K. 1973. The structure of bivariate Poisson distribution. *Kodai Math. Sem. Rep.* 25: 246-256.
18. Kemp, A. W. 1997. Characterization of a discrete normal distribution. *Journal of Statistical Planning and Inference* 63: 223-229.
19. Kemp, A. W. 2004. Classes of discrete lifetime distributions. *Communications in Statistics-Theory and Methods* 33: 3069-3093.
20. Klenke, A., Mattner, L. 2010. Stochastic ordering of classical discrete distributions. *Advances in Applied Probability* 42: 392-410.
21. Kocherlakota, S. and Kocherlakota, K. 1992. *Bivariate Discrete Distributions*, Marsell Dekker, Inc., New York.
22. Koutras, M. V. 1997. Waiting time distributions associated with runs of fixed length in two-state Markov chains. *Annals of the Institute of Statistical Mathematics*, 49, 123-139.
23. Koutras, M.V. and Eryilmaz, S. 2017. Compound geometric distribution of order k. *Methodology and Computing in Applied Probability* 19: 377-393.
24. Li, Z.H. and Kong, X.B. 2007. Life behavior of δ-shock model. *Statistics and Probability Letters* 77: 577-587.
25. Lian, Z.T, Liu, L. M and Neuts, M.F. 2005. A discrete-time model for common lifetime inventory systems. *Mathematics of Operations Research* 30: 718-732.

26. Louit, D., Pascual, R., Banjevic, D. and Jardine, A.K.S. 2011. Optimization models for critical spare parts inventories-a reliability approach. *Journal of the Operational Research Society* 62: 992-1004.
27. Mallor, F., and Omey, E. 2001. Shocks, runs and random sums. *Journal of Applied Probability* 38: 438-448.
28. Marshall, A. W., and Olkin, I. 1985. A family of bivariate distributions generated by the bivariate Bernoulli distribution. *Journal of the American Statistical Association* 80: 332-338.
29. Muselli, M. 1996. Simple expressions for success run distributions in Bernoulli trials. *Statistics and Probability Letters* 31: 121-128.
30. Nakagawa, T. and Osaki, S. 1975. The discrete Weibull distribution. *IEEE Transactions on Reliability* 24: 300 - 301.
31. Pervez, Z., Shah, S.F., Hussain, Z. and Philippou, A.N. 2017. An application of negative binomial distribution of order k in sensitive surveys. *Communications in Statistics - Simulation and Computation* 46: 6654-6660.
32. Philippou, A. N. and Georghiou, C. 1989. Convolutions of Fibonacci type polynomials of order k and the negative binomial distributions of the same order. *The Fibonacci Quarterly* 27: 209-216.
33. Philippou, A.N, Georghiou, C. and Philippou, G.N. 1983. A generalized geometric distribution and some of its properties. *Statistics and Probability Letters* 1:171-175.
34. Roy, D. 2004. Discrete Rayleigh distribution. *IEEE Transactions on Reliability* 53: 255-260.
35. Samanta, S.K., Gupta, U.C. and Chaudhry, M.L. 2009. Analysis of stationary discrete-time GI/D-MSP/1 queue with finite and infinite buffers. *4OR-A Quarterly Journal of Operations Research* 7: 337-361.
36. Teugels, J.L. and Vynckier, P. 1996. The structure distribution in a mixed Poisson process. *Journal of Applied Mathematics and Stochastic Analysis* 9: 489-496.
37. Vandeput, N. 2020. *Inventory optimization*. Walter de Gruyter GmbH, Berlin/Boston.
38. Wang, Y.H. 1993. On the number of successes in independent trials. *Statistica Sinica* 3: 295-312.
39. Zhao, X. and Nakagawa, T. 2018. *Advanced maintenance policies for shock and damage models*. Springer Series in Reliability Engineering.

3 Markov Chains

Under the assumption of the classical Bernoulli trials, the outcome of the next trial is not influenced by the outcome of the previous trial. However, in some real-life applications, successive outcomes might be dependent. For example, the outcomes of the successive tests applied to a certain equipment might be dependent. Such a dependence may be modeled by Markov chains (Martin 2004), (Martin 2008). The states (as either defective or nondefective) of the items produced by a production process can be modeled by a Markov chain (Bebbington et al. 2003). Markov chains have also applied for modeling wind speed series by defining an interval for each wind speed state. That is, wind speed series is transformed into a series of wind speed states with appropriately defined wind speed intervals (Song et al. 2011), (Zheng et al. 2020). Markov chains are also useful in software testing, see, e.g. (Whittaker and Thomason 1994), (Prowell and Poore 2004). There are numerous applications of Markov chains in many fields including finance, psychology, and biology.

Markov chains are useful not only for modeling dependence between successive trials but also for solving various stochastic problems by imbedding them into a properly defined Markov chain. One of the most important achievements in this context is the Markov-chain imbedding technique which has been successfully applied for investigating distributional properties of runs and patterns (Fu and Lou 2003).

Let us begin with the following example to define Markov chains. A certain production process is stopped and declared to be out-of-control whenever two successive plotting statistics fall beyond the fixed control limits LCL and UCL. Let X_n denote the cumulative number of successive plotting statistics beyond the control limits at the nth sampling. Then, $X_n \in \{0,1,2\}$ for $n \geq 0$ with $P\{X_0 = 0\} = 1$. Define a random variable ξ_n as $\xi_n = 1$ if the plotting statistic obtained from the nth sampling is beyond the control limits, and $\xi_n = 0$ if the plotting statistic falls into the control limits. Assume that the random variables $\xi_1, \xi_2, \ldots$ are independent. That is, the successive plotting statistics are obtained independently. To make the definition of the stochastic process $\{X_n\}_{n \geq 0}$ more clear, consider the realization 0001010011 of the sequence $\xi_1, \xi_2, \ldots$. Then,

$$X_1 = 0, X_2 = 0, X_3 = 0, X_4 = 1, X_5 = 0,$$
$$X_6 = 1, X_7 = 0, X_8 = 0, X_9 = 1, X_{10} = 2.$$

Clearly,

$$X_n = \begin{cases} 0, & \text{if } \xi_n = 0 \\ X_{n-1} + 1, & \text{if } \xi_n = 1, \end{cases} \tag{3.1}$$

for $n \geq 1$ with $P\{X_0 = 0\} = 1$. Therefore, for $i = 0, 1$,

$$P\{X_{n+1} = j \mid X_n = i\} = \begin{cases} P\{\xi_{n+1} = 1\}, & \text{if } j = i+1 \\ P\{\xi_{n+1} = 0\}, & \text{if } j = 0. \end{cases} \tag{3.2}$$

DOI: 10.1201/9781003160632-3

On the other hand, for $j = i+1$,

$$\begin{aligned}
&P\{X_{n+1} = j \mid X_n = i,...,X_1 = i_1, X_0 = 0\} \\
&= P\{X_{n+1} = i+1 \mid X_n = i,...,X_1 = i_1\} \\
&= P\{\xi_{n+1} = 1 \mid X_n = i,...,X_1 = i_1\} \\
&= P\{\xi_{n+1} = 1\},
\end{aligned}$$

and for $j = 0$,

$$\begin{aligned}
&P\{X_{n+1} = j \mid X_n = i,...,X_1 = i_1, X_0 = 0\} \\
&= P\{X_{n+1} = 0 \mid X_n = i,...,X_1 = i_1\} \\
&= P\{\xi_{n+1} = 0 \mid X_n = i,...,X_1 = i_1\} \\
&= P\{\xi_{n+1} = 0\}.
\end{aligned}$$

Thus, for the stochastic process $\{X_n\}_{n\geq 0}$, we have

$$\begin{aligned}
&P\{X_{n+1} = j \mid X_n = i,...,X_1 = i_1, X_0 = i_0\} \\
&= P\{X_{n+1} = j \mid X_n = i\}
\end{aligned} \tag{3.3}$$

for all time points n and all states $i_0, i_1, ..., i, j$.

A stochastic process $\{X_n\}_{n\geq 0}$ satisfying the property (3.3) is called a Markov chain. In words, when the current state of the process is exactly known, the probability of the future behavior of the process is not altered by the additional knowledge concerned with its past behavior. That is, the past information included by $X_{n-1}, ..., X_1, X_0$ is redundant since X_n contains all information about the future value X_{n+1} of the process.

The probability $P\{X_{n+1} = j \mid X_n = i\}$ denotes the conditional probability of being in state j at time $n+1$ given that the process is in state i at time n and is called transition probability. If $P\{X_{n+1} = j \mid X_n = i\}$ is constant in time n, i.e. independent of n, then the corresponding Markov chain is known to be homogeneous. The matrix $\mathbf{P}$ which contains conditional probabilities $p_{ij} = P\{X_{n+1} = j \mid X_n = i\}$ for all states i and j is called transition probability matrix of the Markov chain.

For the Markov chain defined by (3.1),

$$\mathbf{P} = \begin{bmatrix} p_{00} & p_{01} & p_{02} \\ p_{10} & p_{11} & p_{12} \\ p_{20} & p_{21} & p_{22} \end{bmatrix} = \begin{bmatrix} p & 1-p & 0 \\ p & 0 & 1-p \\ 0 & 0 & 1 \end{bmatrix}, \tag{3.4}$$

where $p = P\{\xi_n = 0\}$ and $1-p = P\{\xi_n = 1\}$ for all n. Note that the state $\{2\}$ is absorbing since $p_{22} = 1$. That is, if the process enters the state $\{2\}$, then it remains there forever.

The elements p_{ij} of any transition probability matrix $\mathbf{P}$ satisfy the conditions

$$p_{ij} \geq 0 \text{ and } \sum_j p_{ij} = 1.$$

The probability $P\{X_{n+1} = j \mid X_n = i\}$ represents the one-step transition probability from state i to state j. For a Markov chain, the transition from the state i to the state j may be possible in more than one steps. The well-known Kolmogorov-Chapman equation can be used for computing $p_{ij}^{(m)} = P\{X_{n+m} = j \mid X_n = i\}$ which denotes the transition probability from state i to state j in m steps. Using the Markov property of $\{X_n\}_{n\geq 0}$ along with the law of total probability, one obtains

$$p_{ij}^{(m)} = \sum_k p_{ik} p_{kj}^{(m-1)}. \tag{3.5}$$

If $\mathbf{P}^{(m)}$ denotes the matrix which includes m step transition probabilities corresponding to Markov chain $\{X_n\}_{n\geq 0}$, then, as an immediate consequence of (3.5), for a homogeneous Markov chain, we obtain

$$\mathbf{P}^{(m)} = \mathbf{P}^m.$$

That is, the m step transition probabilities $p_{ij}^{(m)}$ are the entries of the mth power of the matrix $\mathbf{P}$. For the Markov chain defined by (3.1),

$$\mathbf{P}^2 = \begin{bmatrix} p & p(1-p) & (1-p)^2 \\ p^2 & p(1-p) & 1-p \\ 0 & 0 & 1 \end{bmatrix},$$

and

$$\mathbf{P}^3 = \begin{bmatrix} p^2(2-p) & p(1-p) & (1+p)(1-p)^2 \\ p^2 & p^2(1-p) & 1-2p^2+p^3 \\ 0 & 0 & 1 \end{bmatrix}.$$

Thus, we have the following two step and three step transition probabilities:

$$\begin{aligned}
&P\{X_{n+2} = 0 \mid X_n = 0\} = p, \\
&P\{X_{n+2} = 1 \mid X_n = 0\} = p(1-p), \\
&P\{X_{n+2} = 0 \mid X_n = 1\} = p^2, \\
&P\{X_{n+3} = 0 \mid X_n = 0\} = p^2(2-p), \\
&P\{X_{n+3} = 1 \mid X_n = 0\} = p(1-p), \\
&P\{X_{n+3} = 0 \mid X_n = 1\} = p^2.
\end{aligned}$$

For a Markov chain $\{X_n\}_{n\geq 0}$, the joint probability $P\{X_0 = i_0, X_1 = i_1, \ldots, X_n = i_n\}$ can be computed from

$$\begin{aligned}
&P\{X_0 = i_0, X_1 = i_1, \ldots, X_n = i_n\} \\
&= P\{X_n = i_n \mid X_{n-1} = i_{n-1}\} \ldots P\{X_1 = i_1 \mid X_0 = i_0\} P\{X_0 = i_0\}.
\end{aligned}$$

That is, a Markov chain can be completely defined by its one-step transition probabilities and the distribution of the initial state, i.e. the distribution of X_0.

Example 3.1. For the above mentioned production process example, recall that the process is declared to be out-of-control whenever two successive plotting statistics fall beyond the control limits LCL and UCL. Let us find the mean number of samples (or plotting statistics) that needs to be collected before the out-of-control signal. This mean is known to be average run length in statistical quality control literature. Because, the process is stopped whenever the corresponding Markov chain hits the absorbing state "2", the number of samples until two consecutive plotting statistics beyond the control limits is defined by

$$T = \min(n : X_n = 2),$$

where X_n denotes the cumulative number of successive plotting statistics beyond the control limits at the nth sampling. Because $P\{X_0 = 0\} = 1$, $E(T) = E(T \mid X_0 = 0)$. Since at least one step is necessary for the absorption,

$$\begin{aligned} E(T) &= E(T \mid X_0 = 0) \\ &= 1 + E(T \mid X_1 = 0)P\{X_1 = 0 \mid X_0 = 0\} \\ &+ E(T \mid X_1 = 1)P\{X_1 = 1 \mid X_0 = 0\} \\ &+ E(T \mid X_1 = 2)P\{X_1 = 2 \mid X_0 = 0\}. \end{aligned}$$

Clearly,

$$\begin{aligned} P\{X_1 = 0 \mid X_0 = 0\} &= P\{X_1 = 0\} = p, \\ P\{X_1 = 1 \mid X_0 = 0\} &= P\{X_1 = 1\} = 1 - p, \\ P\{X_1 = 2 \mid X_0 = 0\} &= 0. \end{aligned}$$

Thus,

$$E(T) = 1 + v_0 p + v_1(1 - p), \tag{3.6}$$

where

$$v_0 = E(T \mid X_1 = 0),$$

and

$$v_1 = E(T \mid X_1 = 1).$$

The conditional expected values v_0 and v_1 can be computed from

$$\begin{aligned} v_0 &= 1 + E(T \mid X_2 = 0)P\{X_2 = 0 \mid X_1 = 0\} \\ &+ E(T \mid X_2 = 1)P\{X_2 = 1 \mid X_1 = 0\} \\ &+ E(T \mid X_2 = 2)P\{X_2 = 2 \mid X_1 = 0\} \\ &= 1 + v_0 p + v_1(1 - p), \end{aligned}$$

and similarly,

$$v_1 = 1 + v_0 p.$$

By solving the last two equations and then using the values of v_0 and v_1 in (3.6), we obtain

$$E(T) = \frac{2 - p}{(1 - p)^2},$$

where p is the probability that plotting statistic falls into the control limits. ■

Example 3.2. Consider a system which can be in one of the three possible states: perfect functioning (denoted by "2"), partial working (denoted by "1") and complete failure ("0"). The system is nonrepairable and it degrades over time. It is assumed to be initially in state "2". Let X_n denote the state of the system at the of the nth time period such as day, week, month, etc. Assume that the transitions among the states occur according to a homogeneous Markov chain with

$$\mathbf{P} = \begin{bmatrix} 1 & 0 & 0 \\ p_{10} & p_{11} & 0 \\ 0 & p_{21} & p_{22} \end{bmatrix},$$

and $P\{X_0 = 2\} = 1$. The state "0" is absorbing since it corresponds to the complete failure. Then, the lifetime of the system is

$$T = \min(n : X_n = 0).$$

The mean time to failure of the system can be computed as follows. Clearly,

$$E(T) = E(T \mid X_0 = 2).$$

Thus, we have

$$\begin{aligned} v_2 = E(T) &= E(T \mid X_0 = 2) \\ &= 1 + E(T \mid X_1 = 1)P\{X_1 = 1 \mid X_0 = 2\} \\ &+ E(T \mid X_1 = 2)P\{X_1 = 2 \mid X_0 = 2\} \\ &= 1 + v_1 p_{21} + v_2 p_{22}, \end{aligned}$$

where

$$\begin{aligned} v_1 &= E(T \mid X_1 = 1), \\ v_2 &= E(T \mid X_0 = 2) = E(T \mid X_1 = 2). \end{aligned}$$

On the other hand,

$$\begin{aligned} v_1 = E(T \mid X_1 = 1) &= 1 + E(T \mid X_2 = 0)P\{X_2 = 0 \mid X_1 = 1\} \\ &+ E(T \mid X_2 = 1)P\{X_2 = 1 \mid X_1 = 1\} \\ &= 1 + v_1 p_{11} \end{aligned}$$

which yields

$$v_1 = \frac{1}{1 - p_{11}}.$$

Therefore,

$$v_2 = \frac{1}{1 - p_{22}} \left(1 + \frac{p_{21}}{1 - p_{11}} \right) = \frac{1}{p_{21}} \left(1 + \frac{p_{21}}{p_{10}} \right).$$

Assume that the time is measured in days and

$$\mathbf{P} = \begin{bmatrix} 1 & 0 & 0 \\ 0.05 & 0.95 & 0 \\ 0 & 0.04 & 0.96 \end{bmatrix}.$$

The mean time to failure of the system is found to be

$$\nu_2 = \frac{1}{0.04}\left(1 + \frac{0.04}{0.05}\right) = 45 \text{ days.} \quad \blacksquare$$

Below, we give an example on a discrete time queueing system which is one of the important models in operational research and engineering applications. The machines arriving at the maintenance center for repair or the products arriving at the quality control station for inspection can be modeled via queueing systems. The main components of a queueing system are input process, service mechanism, system capacity, and service discipline. The input process provides information about the arrival pattern of the customers, i.e. the distribution of the times between arrivals of the customers. The number of servers within the system, the number of customers getting served at any time, and the service time are typical variables that define the service mechanism. The number of customers that can wait at a time in a queueing system is concerned with the system capacity. Finally, the service discipline defines the priority in accepting customers for the service.

In the case of the well-known continuous time $M/M/1$ system, the input process is Poisson process. That is, the interarrival times between successive arrivals follow exponential distribution. The service time has also an exponential distribution, and there is a single server in the system. The following example includes discrete time analogue of the $M/M/1$ system. Namely, the $Geo/Geo/1$ system which admits Binomial process as an input process, i.e. the times between arrivals follow geometric distribution. The second Geo in the notation $Geo/Geo/1$ implies that the service time distribution is also geometric.

Example 3.3. Recall from Example 2.3 that the failed machines arrive at the maintenance center for repair at time marks. Assume that no more than one machine may arrive at a given time mark and the arrivals occur independently. The probability of arrival of a machine at a time mark is p and the probability of no arrival at a time mark is $1-p$. Then, the distribution of the interarrival time between successive arrivals follows geometric distribution with mean $\frac{1}{p}$. The failed machines are served in their order of arrival and the service time distribution is also geometric with mean $\frac{1}{d}$. One of the most important indices in the analysis of queueing systems is the probability distribution of the queue length when the system is in equilibrium. Let

$$\pi_i = \lim_{n \longrightarrow \infty} P\{X_n = i\}$$

be the probability of having a queue length i when the system is in equilibrium, $i = 0, 1, 2, ...$, where X_n denotes the number of customers (machines) in the system at time n. X_n is a Markov chain defined on the states $\{0,1,2,...\}$. Clearly,

$$P\{X_{n+1}=0 \mid X_n=0\}=1-p,$$
$$P\{X_{n+1}=1 \mid X_n=0\}=p,$$

and

$$\begin{aligned} P\{X_{n+1}=i-1 \mid X_n=i\} &= (1-p)d, \\ P\{X_{n+1}=i \mid X_n=i\} &= pd+(1-p)(1-d), \\ P\{X_{n+1}=i+1 \mid X_n=i\} &= p(1-d), \end{aligned}$$

for $i \geq 1$ (see, e.g. Alfa 2002).

Thus, the corresponding transition probability matrix is

$$\mathbf{P}=\begin{bmatrix} 1-p & p & 0 & 0 & \cdots & 0 \\ (1-p)d & pd+(1-p)(1-d) & p(1-d) & 0 & & \\ 0 & (1-p)d & pd+(1-p)(1-d) & p(1-d) & & \\ & & \ddots & \ddots & & \end{bmatrix}.$$

If the Markov chain is positive recurrent, then the limiting distribution

$$\pi_i=\lim_{n\longrightarrow\infty} P\{X_n=i\}$$

is obtained from

$$\pi_j=\sum_{k\geq 0}\pi_k p_{kj}, \quad j=0,1,2,...$$
$$\sum_{k\geq 0}\pi_k=1,$$

where p_{kj} are the entries of the matrix $\mathbf{P}$. Thus, we have

$$\begin{aligned} \pi_0 &= \pi_0(1-p)+\pi_1(1-p)d, \\ \pi_1 &= \pi_0 p+\pi_1[pd+(1-p)(1-d)]+\pi_2(1-p)d, \\ \pi_i &= \pi_{i-1}p(1-d)+\pi_i[pd+(1-p)(1-d)]+\pi_{i+1}(1-p)d,\ i\geq 2. \end{aligned}$$

After solving these equations, the probability of having a queue length i when the system is in equilibrium is obtained as

$$\pi_0=1-\rho,$$

and

$$\pi_i=(1-\rho)\frac{p}{(1-p)d}\left[\frac{p(1-d)}{d(1-p)}\right]^{i-1},$$

for $i \geq 1$ (see, e.g. Alfa 2002), where $\rho = \frac{p}{d}, p < d$. Clearly, $\pi_0 = 1 - \frac{p}{d}$ denotes the probability that the system is idle. For an illustration, let $p = 0.1$ and $d = 0.5$. Then,

$$\begin{aligned}
\pi_0 &= 0.8000, \\
\pi_1 &= 0.1778, \\
\pi_2 &= 0.0198, \\
\pi_3 &= 0.0022.
\end{aligned}$$

Example 3.4. Consider a system that is subject to a sequence of randomly occurring shocks at periods $t = 1, 2, \ldots$. Each shock causes a damage which is represented by a random variable Y, $P\{Y > 0\} = 1$. Assume that the shocks occur according to a Markov chain. That is, if I_t denotes the shock occurrence indicator such that

$$I_t = \begin{cases} 1, & \text{if a shock occurs in period } t \\ 0 & \text{otherwise,} \end{cases}$$

then the sequence $I_1, I_2, \ldots$ forms a Markov chain with given transition probability matrix

$$\mathbf{P} = \begin{bmatrix} p_{00} & p_{01} \\ p_{10} & p_{11} \end{bmatrix},$$

and initial probabilities $p_0 = P\{I_1 = 0\}$ and $p_1 = P\{I_1 = 1\}$, where $p_0 + p_1 = 1$. The system fails if the cumulative damage caused by the shocks exceeds a given level $k > 0$. (Eryilmaz 2016) represented the failure time of the system as

$$T = \sum_{i=1}^{R} X_i,$$

where X_1 denote the time until the first shock, and X_i is the interarrival time between the $(i-1)$st and ith shocks for $i > 1$, and

$$R = \min\{n : Y_1 + \ldots + Y_n \geq k\}.$$

The damages $Y_1, Y_2, \ldots$ caused by the shocks over the periods are assumed to be independent and identically distributed with $\bar{F}(y) = P\{Y_i > y\}$. As shown by (Eryilmaz 2016),

$$P\{X_1 = x\} = \begin{cases} p_1, & \text{if } x = 1 \\ p_{01} p_{00}^{x-2} p_0, & \text{if } x > 1, \end{cases}$$

and for $i > 1$,

$$P\{X_i = x\} = \begin{cases} p_{11}, & \text{if } x = 1 \\ p_{01} p_{00}^{x-2} p_{10}, & \text{if } x > 1. \end{cases}$$

If the sequences $\{I_t\}_{t \geq 1}$ and $Y_1, Y_2, \ldots$ are independent, then the mean time to failure of the system is

$$\begin{aligned}
E(T) = p_1 &+ \frac{p_0 p_{01}(2 - p_{00})}{(1 - p_{00})^2} \\
&+ \left(p_{11} + \frac{p_{01} p_{10}(2 - p_{00})}{(1 - p_{00})^2}\right) [E(R) - 1],
\end{aligned}$$

where

$$E(R) = \sum_{n=1}^{\infty} n\left[\bar{F}_n(k) - \bar{F}_{n-1}(k)\right].$$

Example 3.5. Consider a Markov chain $\{X_i\}_{i\geq 1}$ defined by the following transition probabilities:

$$P\{X_{i+1} = k \mid X_i = l\} = \binom{l}{k}(1-p)^k p^{l-k},$$

for $0 < p < 1$, $l \geq k$ and $k = 0, 1, ..., m$. The initial probability distribution is given by

$$P\{X_1 = k\} = \binom{m}{k}(1-p)^k p^{m-k},$$

$k = 0, 1, ..., m$. Such a Markov chain is useful to study the lifetime of a certain random binary sequence (Surikov 2010). Define

$$V = \min(t : X_t = 0),$$

and

$$\tau = V - 1.$$

(Eryilmaz 2011) obtained the following expression for the expected value of the random variable τ :

$$E(\tau) = v_m - 1,$$

where v_m can be obtained recursively from

$$v_j = \frac{1}{1-(1-p)^j}\left[1 + \sum_{i=1}^{j-1} v_i \binom{j}{i}(1-p)^i p^{j-i}\right],$$

for $j = 2, ..., m$ with $v_1 = \frac{1}{p}$.

3.1 MARKOV BINOMIAL MODEL

Consider a homogeneous $\{0,1\}$-valued Markov chain $X_1, X_2, ...$ whose transition probability matrix is

$$\mathbf{P} = \begin{bmatrix} p_{00} & p_{01} \\ p_{10} & p_{11} \end{bmatrix}, \tag{3.7}$$

and initial probabilities are $p_0 = P\{X_1 = 0\}$, $p_1 = P\{X_1 = 1\} = 1 - p_0$. Let S_n be the number of successes (ones) in $X_1, X_2, ..., X_n$. Depending on the initial trial which can be either success or failure, the probability mass function of the random variable S_n is found to be

$$P\{S_n = l\} = \sum_{r=1}^{l} \binom{l-1}{r-1} g(n, r, l), \tag{3.8}$$

where

$$g(n,r,l) = \sum_{t=0}^{1}\sum_{s=0}^{1}\binom{n-l-1}{r-t-s} p_{11}^{l-r} p_{01}^{r-t} p_{10}^{r-s} p_{00}^{n-l-r+t+s-1} p_t,$$

and $g(n,1,n) = p_1 p_{11}^{n-1}$.

The random variable S_n whose probability mass function is given by (3.8) is said to have a Markov-Binomial distribution (Gabriel 1959, Viveros et al. 1994). For an illustration, let us compute $P\{S_n = l\}$ when $n = 4$ and $l = 3$ using (3.8):

$$P\{S_4 = 3\} = g(4,1,3) + 2g(4,2,3) + g(4,3,3),$$

where

$$g(4,1,3) = p_0 p_{01} p_{11}^2 + p_1 p_{10} p_{11}^2,$$

$$g(4,2,3) = p_1 p_{11} p_{01} p_{10},$$

and

$$g(4,3,3) = 0.$$

Thus, we have

$$P\{S_4 = 3\} = p_0 p_{01} p_{11}^2 + p_1 p_{10} p_{11}^2 + 2p_1 p_{11} p_{01} p_{10}.$$

To verify the equation, the probability $P\{S_4 = 3\}$ can also be computed by writing all possible binary sequences satisfying $\{S_4 = 3\}$. Clearly,

$$\begin{aligned}
P\{S_4 = 3\} &= P\{1110\} + P\{0111\} + P\{1101\} + P\{1011\} \\
&= p_1 p_{11}^2 p_{10} + p_0 p_{01} p_{11}^2 + p_1 p_{11} p_{10} p_{01} + p_1 p_{10} p_{01} p_{11} \\
&= p_0 p_{01} p_{11}^2 + p_1 p_{10} p_{11}^2 + 2p_1 p_{11} p_{01} p_{10}.
\end{aligned}$$

To find the mean number of successes in n Markov dependent trials, let us represent S_n as $S_n = \sum_{i=1}^{n} X_i$. The probability of having a success at the ith trial can be computed from

$$\begin{aligned}
P\{X_i = 1\} = P\{X_i = 1 \mid X_{i-1} = 1\} P\{X_{i-1} = 1\} \\
+ P\{X_i = 1 \mid X_{i-1} = 0\} P\{X_{i-1} = 0\},
\end{aligned} \tag{3.9}$$

for $i \geq 2$ with $p_1 = P\{X_1 = 1\}$. For $p(i) = P\{X_i = 1\}$, the equation (3.9) can be written as a non-homogeneous difference equation in the following form. For $i \geq 2$,

$$p(i) - (p_{11} - p_{01}) p(i-1) = p_{01}, \tag{3.10}$$

with initial condition $p(1) = p_1$. Solving (3.10), one obtains

$$p(i) = \left[p_1 - \frac{p_{01}}{1-(p_{11}-p_{01})}\right](p_{11}-p_{01})^{i-1} + \frac{p_{01}}{1-(p_{11}-p_{01})},$$

for $i \geq 1$. Thus, the mean number of successes in n Markov dependent trials is found to be

$$\begin{aligned} E(S_n) &= \sum_{i=1}^{n} E(X_i) = \sum_{i=1}^{n} p(i) \\ &= \frac{np_{01}}{1-(p_{11}-p_{01})} \\ &+ \frac{p_1 - p_1(p_{11}-p_{01}) - p_{01}}{[1-(p_{11}-p_{01})]^2} [1-(p_{11}-p_{01})^n]. \end{aligned}$$

For large n, an approximate value for $E(S_n)$ can be obtained using the limiting probability distribution of the Markov chain $X_1, X_2, \ldots$. Because the transition probability matrix (3.7) is regular for $0 < p_{01}, p_{11} < 1$, the limiting distribution (π_0, π_1) of the Markov chain $X_1, X_2, \ldots$ is obtained from

$$\begin{aligned} \pi_0 &= \pi_0 p_{00} + \pi_1 p_{10}, \\ \pi_1 &= \pi_0 p_{01} + \pi_1 p_{11}, \end{aligned}$$

with $\pi_0 + \pi_1 = 1$. The probability of success or the probability of the Markov chain is in state "1" in the long run is obtained as

$$\pi_1 = \frac{p_{01}}{1-(p_{11}-p_{01})}.$$

Thus, for large n,

$$E(S_n) \sim \frac{np_{01}}{1-(p_{11}-p_{01})}.$$

We compute $E(S_n)$ for selected values of the parameters:

$$\begin{aligned} n = 50, p_{01} = 0.3, p_{11} = 0.9 &\Longrightarrow E(S_n) = 37.5, \\ n = 50, p_{01} = 0.4, p_{11} = 0.9 &\Longrightarrow E(S_n) = 40, \\ n = 100, p_{01} = 0.4, p_{11} = 0.9 &\Longrightarrow E(S_n) = 80. \end{aligned}$$

3.2 SIMULATION OF A MARKOV CHAIN

Let us first consider a homogeneous Markov chain with transition probability matrix

$$\mathbf{P} = \begin{bmatrix} p_{00} & p_{01} \\ p_{10} & p_{11} \end{bmatrix},$$

and initial probabilities $p_0 = P\{X_1 = 0\}$, $p_1 = P\{X_1 = 1\} = 1 - p_0$. For given $\mathbf{P}$ and p_0, a sequence $x_1, x_2, \ldots, x_n$ can be generated by using the following algorithm:

►Generate u from unit uniform distribution
►If $u \leq p_0$, then $x_1 = 0$ else $x_1 = 1$
►For $k = 2, \ldots, n$,

►Generate u_k from unit uniform distribution,
►If $x_{k-1}=0$ and $u_k \leq p_{00}$, then $x_k=0$,
►If $x_{k-1}=0$ and $u_k > p_{00}$, then $x_k=1$,
►If $x_{k-1}=1$ and $u_k \leq p_{10}$, then $x_k=0$,
►If $x_{k-1}=1$ and $u_k > p_{10}$, then $x_k=1$.

From a given Markov chain of n elements, the entries of the transition probability matrix $\mathbf{P}$ can be estimated from

$$\hat{p}_{00} = \frac{n_{00}}{n_{00}+n_{01}},$$

$$\hat{p}_{10} = \frac{n_{10}}{n_{10}+n_{11}},$$

where n_{ij} is the number of transitions from state i to state j in $x_1,...,x_n$, $i,j \in \{0,1\}$.

Example 3.6. A sequence of binary trials

$$11000110011111100111$$

have been generated from a two state Markov chain with transition probability matrix

$$\mathbf{P} = \begin{bmatrix} 0.6 & 0.4 \\ 0.2 & 0.8 \end{bmatrix},$$

and initial probabilities $p_0 = P\{X_1=0\} = 0.5$, $p_1 = P\{X_1=1\} = 0.5$. Based on these $n=20$ trials, the estimates of the transition probabilities are found to be

$$\hat{p}_{00} = \frac{4}{7} \simeq 0.57, \ \hat{p}_{01} = \frac{3}{7} \simeq 0.43,$$

$$\hat{p}_{10} = \frac{3}{12} = 0.25, \ \hat{p}_{11} = \frac{9}{12} = 0.75. \quad \blacksquare$$

Consider a Markov chain defined on states $\{1,2,...,r\}$ with transition probability matrix

$$\mathbf{P} = \begin{bmatrix} p_{11} & p_{12} & \cdots & p_{1r} \\ p_{21} & p_{22} & \cdots & p_{2r} \\ \vdots & \vdots & \ddots & \vdots \\ p_{r1} & p_{r2} & \cdots & p_{rr} \end{bmatrix},$$

and $p_i = P\{X_1 = i\}, i=1,2,...,r$. For given $\mathbf{P}$ and p_is, a sequence $x_1,x_2,...,x_n$ can be generated by using the following algorithm:

►Generate u from unit uniform distribution
►If $u \leq p_1$, then $x_1 = 1$, and if $p_1+...+p_{j-1} < u \leq p_1+...+p_j$, then $x_1 = j$, $j = 2,...,r$
►For $k=2,...,n$,
►Generate u_k from unit uniform distribution,
►If $x_{k-1}=i$ and $p_{i1}+p_{i2}+...+p_{i.j-1} < u_k \leq p_{i1}+p_{i2}+...+p_{i.j}$, then $x_k = j, j=1,...,r$.

For a Markov chain with states $\{1,2,...,r\}$, the transition probability from state i to state j can be estimated from

$$\hat{p}_{ij} = \frac{n_{ij}}{\sum_{j=1}^{r} n_{ij}},$$

where n_{ij} is the number of transitions from state i to state j in $x_1,...,x_n$.

Example 3.7. Markov chains have been found to be useful for modeling wind speed in a certain location. Although wind speed measurements are continuous values, they have been modeled via Markov chains by discretization. Suppose we have a series of hourly wind speed measurements at hand. Our aim is to model wind speed by Markov chains. To this end, the wind speed series is converted to a sequence which is defined on the state space $\{1,2,...,r\}$. The wind speed v is in state j if $v \in [j_l, j_u)$, $j = 1,2,...,r-1$ and in state r if $v \in [r_l, \infty)$. The choice of the number of states and the intervals depend on some factors such as wind speed variation and wind turbine specifications. For a simulated hourly wind speed data over a year, assume that the wind speed v (m/s) is in state i if $v \in [2i-2, 2i)$ for $i = 1,...,6$ and in state 7 if $v \in [12, \infty)$. In this case, $r = 7$. For the simulated data, the corresponding transition probability matrix is estimated as

$$\mathbf{P} = \begin{bmatrix} 0.8204 & 0.1791 & 0.0005 & 0 & 0 & 0 & 0 \\ 0.1220 & 0.7112 & 0.1638 & 0.0030 & 0 & 0 & 0 \\ 0.0005 & 0.2396 & 0.6202 & 0.1354 & 0.0043 & 0 & 0 \\ 0 & 0.0089 & 0.2735 & 0.5706 & 0.1392 & 0.0078 & 0 \\ 0 & 0 & 0.0298 & 0.3424 & 0.4988 & 0.1216 & 0.0074 \\ 0 & 0 & 0 & 0.0744 & 0.3967 & 0.3884 & 0.1405 \\ 0 & 0 & 0 & 0 & 0.0789 & 0.4474 & 0.4737 \end{bmatrix}.$$

The limiting distribution of the corresponding Markov chain arises as the solution of the following equations:

$$\pi_j = \sum_{k=1}^{r} \pi_k p_{kj}, \quad j = 1,2,...,r,$$
$$\sum_{k=1}^{r} \pi_k = 1. \tag{3.11}$$

According to (3.11), the limiting distribution of the Markov chain is found to be

$$\pi = \begin{pmatrix} 0.2362 & 0.3467 & 0.2371 & 0.1161 & 0.04585 & 0.01375 & 0.0043 \end{pmatrix}.$$

Based on the limiting distribution, we can say that the wind speed is in state 2 with probability 0.3467 in the long run.

Example 3.8. Consider the linear consecutive-k-out-of-n:F system that has been previously considered in Example 1.10. Assume now that the components are dependent in a Markovian fashion. That is,

$$\begin{aligned}
P\{X_i = 1 \mid X_{i-1} = 1\} &= p_{11}, \\
P\{X_i = 1 \mid X_{i-1} = 0\} &= p_{01}, \\
P\{X_i = 0 \mid X_{i-1} = 1\} &= p_{10}, \\
P\{X_i = 0 \mid X_{i-1} = 0\} &= p_{00},
\end{aligned}$$

for $i = 2, 3, ..., n$, and $p_0 = P\{X_1 = 0\}$, $p_1 = P\{X_1 = 1\} = 1 - p_0$. The methodology described in Example 1.10 can be applied to simulate the reliability of the system. The only difference is that the states of the components must be generated from a given Markov chain. For an illustration, we approximate the reliability of the system for selected values of n and k when $p_{00} = 0.5, p_{11} = 0.98$ and $p_1 = 0.9$:

$$\begin{aligned}
n = 20, k = 2 &\Longrightarrow R = 0.8012, \\
n = 20, k = 3 &\Longrightarrow R = 0.8994, \\
n = 30, k = 3 &\Longrightarrow R = 0.8499.
\end{aligned}$$

Example 3.9. As an extension of the linear consecutive-k-out-of-n:F system model, consider the linear consecutive-weighted-k-out-of-n:F system. The linear consecutive-weighted-k-out-of-n:F system consists of n components, where component i has positive integer weight ω_i, such that the system fails iff the total weight of the consecutively failed components is at least k (Wu and Chen 1994). In this model, k represents the minimum total weight that causes system failure and hence k might be greater than n. Let S_i denote the state of the ith component such that $S_i = 1$ if the ith component has failed and $S_i = 0$ if the ith component is working, $i = 1, 2, ..., n$. Define

$$Y_i = \omega_i S_i,\ i = 1, 2, ..., n,$$

and for $i = 0, 1, ..., n-1$,

$$\theta_{i+1} = \begin{cases} \theta_i + Y_{i+1}, & \text{if } Y_{i+1} \neq 0 \\ 0, & \text{if } Y_{i+1} = 0, \end{cases}$$

with $\theta_0 = 0$. Assume that the components are independent and the reliability of the ith component is

$$p_i = P\{S_i = 0\},$$

$i = 1, 2, ..., n$. As stated by (Eryilmaz and Tutuncu 2009), the sequence $\{\theta_i\}_{i \geq 1}$ is Markovian with the following transition probabilities:

$$P\{\theta_{i+1} = m \mid \theta_i = l\} = \begin{cases} p_{i+1}, & \text{if } m = 0 \\ 1 - p_{i+1}, & \text{if } m = l + \omega_{i+1}, \end{cases}$$

for $l \geq 0$, and $i = 1, 2, ..., n-1$. Using Markovian property of $\{\theta_i\}_{i \geq 1}$, (Eryilmaz and Tutuncu 2009) proposed the following equation to compute the reliability of the system:

$$\begin{aligned}
R &= P\{\theta_1 < k, ..., \theta_n < k\} \\
&= \sum_{i_1=0}^{k-1}\sum_{i_2=0}^{k-1}\cdots\sum_{i_n=0}^{k-1} P\{\theta_1 = i_1, ..., \theta_n = i_n\} \\
&= \sum_{i_1=0}^{k-1}\sum_{i_2=0}^{k-1}\cdots\sum_{i_n=0}^{k-1}\prod_{m=1}^{n-1} P\{\theta_{m+1} = i_{m+1} \mid \theta_m = i_m\} P\{\theta_1 = i_1\}.
\end{aligned}$$

To approximate the value of R by the method of Monte Carlo simulation, the following algorithm can be used. The inputs of the algorithm are

$$\begin{aligned}
&n, k \\
&p_1, p_2, ..., p_n, \\
&\omega_1, \omega_2, ..., \omega_n.
\end{aligned}$$

We first generate a sequence components' states $\mathbf{S}^{(1)} = (S_1^{(1)}, S_2^{(1)}, ..., S_n^{(1)})$ and then compute the value of $\max(\theta_1^{(1)}, ..., \theta_n^{(1)})$ based on $\mathbf{S}^{(1)}$. A random vector of binary sequence $\mathbf{S}^{(2)} = (S_1^{(2)}, S_2^{(2)}, ..., S_n^{(2)})$ is generated and $\max(\theta_1^{(2)}, ..., \theta_n^{(2)})$ is computed based on $\mathbf{S}^{(2)}$. The same procedure is repeated m times. For sufficiently large m, the reliability of the system can be approximated by

$$R \simeq \frac{1}{m}\sum_{i=1}^{m} I\left(\max(\theta_1^{(i)}, ..., \theta_n^{(i)}) < k\right).$$

3.3 TESTING THE ORDER OF MARKOV DEPENDENCE

In the case when the future behavior of the process $\{X_n\}_{n\geq 0}$ depends not only on the current state but also on the last r values of the process, the process $\{X_n\}_{n\geq 0}$ is called Markov chain of order r. In this case, we have

$$\begin{aligned}
&P\{X_{n+1} = j \mid X_n = i_n, ..., X_1 = i_1, X_0 = i_0\} \\
&= P\{X_{n+1} = j \mid X_n = i_n, ..., X_{n-r+1} = i_{n-r+1}\},
\end{aligned}$$

for all time points n and all states $i_0, i_1, ..., i_n, j$. Such a Markov chain is studied via the transition probabilities

$$P\{X_{n+1} = j \mid X_n = i_n, ..., X_{n-r+1} = i_{n-r+1}\},$$

and initial probabilities

$$P\{X_0 = i_0, X_1 = i_1, ..., X_{r-1} = i_{r-1}\},$$

for $r \geq 1$, where

$$\sum_j P\{X_{n+1} = j \mid X_n = i_n, ..., X_{n-r+1} = i_{n-r+1}\} = 1$$

for all $i_{n-r+1}, ..., i_n$. Clearly, when $r = 1$ we have the usual Markov chain.

Modeling a sequential data with higher order Markov chains might be more feasible in certain problem setups. In this context, the determination of the value of r is of special importance in decision making processes. The method of likelihood ratio has been used to test if the order of the Markov chain $\{X_n\}_{n\geq 0}$ is $r-1$ or r, i.e. H_0 : Order is $r-1$ against H_1 : Order is r. Let $X_1, X_2, ..., X_n$ be sample from $\{X_n\}_{n\geq 0}$. Then, for testing H_0 against H_1, the likelihood ratio test is given by (Menendez et al. 2011)

$$LR = -2\left(\sum_{a_2,\dots,a_{r+1}} n_{a_2,\dots,a_r,a_{r+1}} \log \frac{n_{a_2,\dots,a_r,a_{r+1}}}{n_{a_2,\dots,a_r}} - \sum_{a_1,\dots,a_{r+1}} n_{a_1,\dots,a_r,a_{r+1}} \log \frac{n_{a_1,\dots,a_r,a_{r+1}}}{n_{a_1,\dots,a_r}}\right), \quad (3.12)$$

for $r > 1$, where

$$n_{a_1,\dots,a_r,a_{r+1}} = \sum_{k=1}^{n-r} I(X_k = a_1, \dots, X_{k+r-1} = a_r, X_{k+r} = a_{r+1}),$$

and $I(A) = 1$ if A occurs and $I(A) = 0$ if A does not occur. If $r = 1$, then

$$LR = -2\left(\sum_{a_1} n_{a_1} \log \frac{n_{a_1}}{n} - \sum_{a_1,a_2} n_{a_1,a_2} \log \frac{n_{a_1,a_2}}{n_{a_1}}\right), \quad (3.13)$$

where

$$n_{a_1} = \sum_{k=1}^{n} I\{X_k = a_1\},$$
$$n_{a_1,a_2} = \sum_{k=1}^{n-1} I\{X_k = a_1, X_{k+1} = a_2\}.$$

It should be noted that in the special case when $r = 1$, the test is of the null hypothesis that $X_1, X_2, ...$ consists of independent and identical trials against the alternative hypothesis that trials are from a first-order Markov chain. The asymptotic distribution of (3.12) and (3.13) is chi-squared with $(m^r - m^{r-1})(m-1)$ degrees of freedom under H_0 (Hoel 1954), where m is the number of states for the Markov chain under concern.

Example 3.10. For a simulated binary data of size $n = 30$, to find the order of the sequence, we apply the likelihood ratio test. The simulated data is

000111111111111111111011111111.

We first test H_0 : Order is 0 versus H_1 : Order is 1. From (3.13), the value of the test statistic is found to be $LR = 7.6570$. Because $\chi^2_{1,0.05} = 3.8410$, the null hypothesis is rejected. Next, we test H_0 : Order is 1 versus H_1 : Order is 2. From (3.12), the test statistic is computed as $LR = 2.9462$. In this case, the critical value is $\chi^2_{2,0.05} = 5.99$. Therefore, H_0 cannot be rejected. Thus, the data under consideration has order 1 and the estimate of the transition probability matrix is

$$\hat{\mathbf{P}} = \begin{bmatrix} 0.5 & 0.5 \\ 0.04 & 0.96 \end{bmatrix}.$$

Example 3.11. Assume that the outcomes of the successive tests applied to a certain equipment form a second order Markov chain with transition probabilities

$$P\{X_{n+1} = 1 \mid X_n = 0, X_{n-1} = 0\} = \frac{18}{87},$$
$$P\{X_{n+1} = 1 \mid X_n = 1, X_{n-1} = 0\} = \frac{8}{21},$$
$$P\{X_{n+1} = 1 \mid X_n = 0, X_{n-1} = 1\} = \frac{1}{3},$$
$$P\{X_{n+1} = 1 \mid X_n = 1, X_{n-1} = 1\} = \frac{19}{21},$$

for $n > 1$. The initial probabilities are given by

$$P\{X_2 = 0, X_1 = 0\} = \frac{21}{100},$$
$$P\{X_2 = 1, X_1 = 0\} = P\{X_2 = 0, X_1 = 1\} = \frac{9}{100},$$
$$P\{X_2 = 1, X_1 = 1\} = \frac{61}{100},$$

where X_n denotes the outcome of the nth test such that

$$X_n = \begin{cases} 1, & \text{if the } n\text{th test is successful} \\ 0, & \text{if the } n\text{th test is failure} \end{cases}$$

Let us find the probability that the first failure occurs on the fourth test. The desired probability is

$$\begin{aligned}
&P\{X_4 = 0, X_3 = 1, X_2 = 1, X_1 = 1\} \\
&= P\{X_4 = 0 \mid X_3 = 1, X_2 = 1, X_1 = 1\} P\{X_3 = 1, X_2 = 1, X_1 = 1\} \\
&= P\{X_4 = 0 \mid X_3 = 1, X_2 = 1\} P\{X_3 = 1 \mid X_2 = 1, X_1 = 1\} \\
&\times P\{X_2 = 1, X_1 = 1\} \\
&= \frac{2}{21}\frac{19}{21}\frac{61}{100} \\
&\simeq 0.0526. \quad \blacksquare
\end{aligned}$$

3.4 MARKOV CHAIN WITH A CHANGE POINT

As it was stated above, if the transition probabilities $P\{X_{n+1} = j \mid X_n = i\}$ are constant in time n, then the corresponding Markov chain is time homogeneous. However, the transition probabilities or equivalently the behavior of the corresponding process may change due to certain causes. Such a situation can be modeled by a Markov chain with change point. For example, as illustrated by (Polansky 2007), the states of a certain industrial machine which are observed over hours can be modeled by a Markov chain with a change point when a new work flow policy is proposed by the management at the plant. Markov chains with a change point have been studied by (Maltsev and Silaev 1992) and (Silaev 1997).

For a Markov chain with a single change point c, the transition probabilities can be defined as

$$P\{X_{n+1}=j \mid X_n=i\}=p_{ij},\ n=1,2,...,c-1,$$

and

$$P\{X_{n+1}=j \mid X_n=i\}=q_{ij},\ n=c,c+1,...$$

Example 3.12. Let $\{X_n\}_{n\geq 1}$ be a two-state Markov chain having a single change point such that

$$P\{X_{n+1}=j \mid X_n=i\}=p_{ij},\ n=1,2,...,c-1,$$

and

$$P\{X_{n+1}=j \mid X_n=i\}=q_{ij},\ n=c,c+1,...,$$

for $i,j\in\{0,1\}$, and $P\{X_1=1\}=p_1$. That is, we have the following transition probability matrices:

$$\mathbf{P}=\begin{bmatrix} p_{00} & p_{01} \\ p_{10} & p_{11} \end{bmatrix},\ \mathbf{Q}=\begin{bmatrix} q_{00} & q_{01} \\ q_{10} & q_{11} \end{bmatrix}.$$

Let W denote the number of trials until the first success ("1") in $\{X_n\}_{n\geq 1}$. Then,

$$P\{W=x\}=\begin{cases} p_1, & \text{if } x=1 \\ p_0 p_{00}^{x-2} p_{01}, & \text{if } 1<x\leq c \\ p_0 p_{00}^{c-1} q_{00}^{x-c-1} q_{01}, & \text{if } x>c. \end{cases}$$

The mean of W can be computed as

$$\begin{aligned} E(W) &= p_1+\sum_{x=2}^{c} x p_0 p_{00}^{x-2} p_{01}+\sum_{x=c+1}^{\infty} x p_0 p_{00}^{c-1} q_{00}^{x-c-1} q_{01} \\ &= p_1+\frac{p_0\left[p_{00}^{c-1}(cp_{00}-1-c)+2-p_{00}\right]}{p_{01}}+p_0 p_{00}^{c-1}\frac{[1+cq_{01}]}{q_{01}}. \end{aligned}$$

For an illustration, let

$$\mathbf{P}=\begin{bmatrix} 0.8 & 0.2 \\ 0.3 & 0.7 \end{bmatrix},\ \mathbf{Q}=\begin{bmatrix} 0.7 & 0.3 \\ 0.5 & 0.5 \end{bmatrix},$$

and $p_1 = 0.5$. Then,

$$c = 5 \Longrightarrow E(W) = 3.1587$$
$$c = 10 \Longrightarrow E(W) = 3.3882$$
$$c = 15 \Longrightarrow E(W) = 3.4633$$
$$c = 50 \Longrightarrow E(W) = 3.5.$$

REFERENCES

1. Alfa, A.S. 2002. Discrete time queues and matrix-analytic methods. *TOP* 10: 147-210.
2. Bebbington, M., Lai, C-D. and Govindaraju, K. 2003. Continuous sampling plans for Markov-dependent production processes under limited inspection capacity. *Mathematical and Computer Modelling* 38: 1137-1145.
3. Eryilmaz, S. 2011. On the lifetime of a random binary sequence. *Discrete Applied Mathematics* 159: 1646-1649.
4. Eryilmaz, S. 2016. Discrete time shock models in a Markovian environment. *IEEE Transactions on Reliability* 65: 141-146.
5. Eryilmaz, S. and Tutuncu, G.Y. 2009. Reliability evaluation of linear consecutive-weighted-k-out-of-n:F system. *Asia-Pacific Journal of Operational Research* 26: 805-816.
6. Fu, J.C. and Lou, W.Y.W., 2003. *Distribution Theory of Runs and Patterns and its Applications, A Finite Markov Chain Imbedding Technique*. World Scientific, Singapore.
7. Gabriel, K.R. 1959. The distribution of the number of successes in a sequence of dependent trials. *Biometrika* 46: 454-460.
8. Hoel, P.G. 1954. A test for Markoff chains. *Biometrika* 41: 430-433.
9. Maltsev, A.A. and Silaev, A.M. 1992. Optimal estimation of the time of change in the characteristics of a Markov chain. *Automat. Remote Control* 53: 52-58.
10. Martin, D.E.K. 2004. Markovian start-up demonstration tests with rejection of units upon observing *d* failures. *European Journal of Operational Research* 155: 474-486.
11. Martin, D.E.K. 2008. Application of auxiliary Markov chains to start-up demonstration tests. *European Journal of Operational Research* 184: 574-583.
12. Menendez, M.L., Pardo, L., Pardo, M.C. and Zografos, K. 2011. Testing the order of Markov dependence in DNA sequences. *Methodology and Computing in Applied Probability* 13: 59-74.
13. Polansky, A.M. 2007. Detecting change-points in Markov chains. *Computational Statistics and Data Analysis* 51: 6013-6026.
14. Prowell, S.J. and Poore, J.H. 2004. Computing system reliability using Markov chain usage models. *The Journal of Systems and Software* 73: 219-225.
15. Silaev, A.M. 1997. Optimal estimation of the parameters of Markovian sequences that change their characteristics at random time instants. *Automat. Remote Control* 58: 1601-1610.
16. Song, Z., Geng, X., Kusiak, A. and Xu, C. 2011. Mining Markov chain transition matrix from wind speed time series data. *Expert Systems with Applications* 38: 10229-10239.
17. Surikov, V.N. 2010. The life time of a random binary sequence (coherent system). *Discrete Mathematics and Applications* 20: 221-230.
18. Viveros, R., Balasubramanian, K. and Balakrishnan, N. 1994. Binomial and negative Binomial analogues under correlated Bernoulli trials. *The American Statistician* 48: 243-247.

19. Whittaker, J.A. and Thomason, M.G. 1994. A Markov chain model for statistical software testing. *IEEE Transactions on Software Testing* 20: 812-824.
20. Wu, J.S and Chen, R.J. 1994. Efficient algorithms for k-out-of-n & consecutive-weighted-k-out-of-n:F system. *IEEE Transactions on Reliability* 43: 650-655.
21. Zheng, R., Zhou, Y. and Zhang, Y. 2020. Optimal preventive maintenance for wind turbines considering the effects of wind speed. *Wind Energy* 23: 1987-2003.

4 Phase-type Distributions

Phase-type distributions are suitable not only for modeling discrete stochastic structures/systems, they are also useful tools for solving problems arising in various fields including reliability engineering, statistical quality control, inventory, and queueing systems. Some contributions in these fields are outlined here and more detail can be found in (Neuts and Meier 1981), (Lian et al. 2005), (Montoro-Cazorla and Perez-Ocon 2010), (Koutras and Rakitsiz 2019), and (Alkaff and Qomarudin 2020).

4.1 DEFINITION AND CHARACTERISTICS

A discrete phase-type distribution is the distribution of the time to absorption in an absorbing Markov chain. Let T denote the number of steps until absorption in a finite Markov chain $\{X_n\}_{n\geq 0}$ with m transient states and one absorbing state. Then, the probability mass function of T has the form

$$P\{T=t\} = \pi \mathbf{Q}^{t-1}\mathbf{u}', \tag{4.1}$$

for $t = 1,2,...,$ and $P\{T=0\} = \pi_0 = 1-\pi\mathbf{e}'$, where $\mathbf{Q}$ denotes the matrix that includes the transition probabilities among the m transient states which are denoted by $\{1,2,...,m\}$, $\mathbf{u}'=(\mathbf{I}-\mathbf{Q})\mathbf{e}'$ is a column vector that includes the transition probabilities from transient states to the absorbing state $\{0\}$, $\mathbf{e}=(1,...,1)_{1\times m}$ and $\mathbf{I}$ denotes $m\times m$ identity matrix. The vector $\pi = (\pi_1,...,\pi_m)$ is the initial transition probability vector with the entry corresponding to the absorption state π_0 removed, where $\pi_i = P\{X_0 = i\}$ and $\sum_{i=0}^{m}\pi_i = 1$ (see, e.g. Neuts 1981). The matrix $\mathbf{Q}$ must satisfy the condition that $\mathbf{I}-\mathbf{Q}$ is nonsingular. We shall use $T\sim PH_m(\pi,\mathbf{Q})$ to represent that the random variable T has a discrete phase-type distribution of order m.

Geometric distribution can be seen as the simplest waiting time distribution that has a phase representation. Indeed, if T has a geometric distribution with probability mass function $P\{T=t\} = p(1-p)^{t-1}, t = 1,2,...m$, then $T\sim PH_1(1,1-p)$. That is, the random variable T can be represented as the time until absorption in a Markov chain with one transient and one absorbing states. Similarly, consider the negative Binomial distribution with the probability mass function

$$P\{T=t\} = \binom{t-1}{r-1}p^r(1-p)^{t-r}, t = r, r+1, ...$$

DOI: 10.1201/9781003160632-4

Then, $T \sim PH_r(\pi, \mathbf{Q})$ with $\pi = (1, 0, ..., 0)$, and

$$\mathbf{Q} = \begin{bmatrix} 1-p & p & \cdots & \cdots & 0 \\ 0 & 1-p & p & \cdots & 0 \\ \vdots & & \ddots & & \vdots \\ 0 & \cdots & \cdots & 0 & 1-p \end{bmatrix}_{r \times r},$$

and

$$\mathbf{u}' = \begin{bmatrix} 0 \\ 0 \\ \vdots \\ p \end{bmatrix}.$$

The cumulative distribution function and the probability generating function of the phase-type random variable T are given respectively by

$$P\{T \leq t\} = 1 - \pi \mathbf{Q}^t \mathbf{e}', \tag{4.2}$$

$$\phi(z) = 1 - \pi \mathbf{e}' + \pi z (\mathbf{I} - z\mathbf{Q})^{-1} \mathbf{u}', \tag{4.3}$$

for $0 < z < 1$.

The mean of the random variable T can be computed from

$$E(T) = \pi (\mathbf{I} - \mathbf{Q})^{-1} \mathbf{e}'. \tag{4.4}$$

The factorial moments of T are computed by successive differentiation of the probability generating function and are given by

$$E(T(T-1)...(T-(k-1))) = k! \pi (\mathbf{I} - \mathbf{Q})^{-k} \mathbf{Q}^{k-1} \mathbf{e}'.$$

In particular, for $k = 2$,

$$E(T(T-1)) = 2\pi (\mathbf{I} - \mathbf{Q})^{-2} \mathbf{Q} \mathbf{e}'.$$

Thus, the variance of T is obtained as

$$Var(T) = 2\pi (\mathbf{I} - \mathbf{Q})^{-2} \mathbf{Q} \mathbf{e}' + \pi (\mathbf{I} - \mathbf{Q})^{-1} \mathbf{e}' - \left[\pi (\mathbf{I} - \mathbf{Q})^{-1} \mathbf{e}'\right]^2.$$

Example 4.1. Consider the production process example previously analyzed in Example 3.1. Recall that the process is declared to be out-of-control whenever two successive plotting statistics fall beyond the control limits LCL and UCL. The number of samples until two consecutive plotting statistics beyond the control limits can be represented as a phase-type random variable. Define a Markov chain with three states. State 1 denoted by $\{1\}$ represents a point within the control limits, State 2 denoted by $\{2\}$ represents a point beyond the control limits, and State 0 (absorbing state) denoted by $\{0\}$ represents two successive points beyond the control limits. Thus, the number of samples until two consecutive points beyond the control limits has a phase-type distribution with representation $T \sim PH_2(\pi, \mathbf{Q})$, where

$\pi = (\pi_1, \pi_2) = (1,0)$ and the matrix $\mathbf{Q}$ includes the transition probabilities among the states $\{1\}$ and $\{2\}$. Manifestly,

$$\mathbf{Q} = \begin{bmatrix} p & 1-p \\ p & 0 \end{bmatrix}, \quad \mathbf{u}' = (\mathbf{I} - \mathbf{Q})\mathbf{e}' = \begin{bmatrix} 0 \\ 1-p \end{bmatrix},$$

where p is the probability that plotting statistic falls into the control limits. Thus, using (4.1) and (4.4), the probability mass function and the mean of T can be computed from

$$P\{T = t\} = (1,0) \begin{bmatrix} p & 1-p \\ p & 0 \end{bmatrix}^{t-1} \begin{bmatrix} 0 \\ 1-p \end{bmatrix},$$

for $t = 2, 3, \ldots$, and

$$\begin{aligned} E(T) &= (1,0) \left(\begin{bmatrix} 1 & 0 \\ 0 & 1 \end{bmatrix} - \begin{bmatrix} p & 1-p \\ p & 0 \end{bmatrix} \right)^{-1} \begin{bmatrix} 1 \\ 1 \end{bmatrix} \\ &= \frac{2-p}{(1-p)^2}. \end{aligned}$$

The variance of T is found to be

$$Var(T) = \frac{p^3 - 5p^2 + 5p}{(1-p)^4}.$$

Example 4.2. Consider a system that consists of n linearly ordered components labeled from 1 to n. Assume that the system fails if there exist at least f failed components or at least k consecutive failed components. The components can be in one of two states either working (1) or failed (0), and during a fixed period of time each component works with probability p. For an illustration, let $n = 5, f = 3$, and $k = 2$. If for example the states of the components are 10101, then the system is an operational state since the total number of failed components is less than $f = 3$ and there are no $k = 2$ consecutively failed components. We aim to compute the working probability (reliability) of the system. Without a Markov chain knowledge, this problem

Table 4.1
All Possible Sequences that Ensure the Functioning of the System

Sequence	Probability	Sequence	Probability
11010	$p^3(1-p)^2$	11110	$p^4(1-p)$
01110	$p^3(1-p)^2$	10111	$p^4(1-p)$
01101	$p^3(1-p)^2$	11011	$p^4(1-p)$
10101	$p^3(1-p)^2$	01111	$p^4(1-p)$
10110	$p^3(1-p)^2$	11101	$p^4(1-p)$
01011	$p^3(1-p)^2$	11111	p^5

can be solved by listing all possible binary sequences of size n if n is not large. In Table 4.1, we list all possible sequences that ensure the functioning of the system when $n = 5$.

From Table 4.1, the reliability of the system is found to be

$$\begin{aligned} R &= 6p^3(1-p)^2 + 5p^4(1-p) + p^5 \\ &= 6p^3 - 7p^4 + 2p^5. \end{aligned}$$

Obviously, this method is not efficient when the system size n is large. In the following, using phase-type distribution, we obtain a matrix-based equation for R for an arbitrary value of n. Let T be the number of trials until getting either $f = 3$ failures or $k = 2$ consecutive failures in a sequence of Bernoulli trials with the probability of success p. Then, the reliability of the system can be expressed as

$$R = P\{T > n\}.$$

Define a bivariate Markov chain such that the first component counts the cumulative number of failed components and the second component keeps the cumulative number of consecutive failures in a sequence of Bernoulli trials. The transient states for the corresponding Markov chain are $(0,0), (1,0), (1,1), (2,0)$, and $(2,1)$. The absorbing states are $(2,2), (3,1)$, and $(3,2)$. The Markov chain is initially in state $(0,0)$ and the matrix that involves transition probabilities among the five transient states is

$$\mathbf{Q} = \begin{bmatrix} p & 0 & 1-p & 0 & 0 \\ 0 & p & 0 & 0 & 1-p \\ 0 & p & 0 & 0 & 0 \\ 0 & 0 & 0 & p & 0 \\ 0 & 0 & 0 & p & 0 \end{bmatrix}.$$

Thus, $T \sim PH_5(\pi, \mathbf{Q})$ and hence the reliability of the system can be computed from

$$R = P\{T > n\} = \pi \mathbf{Q}^n \mathbf{e}',$$

where $\pi = (1,0,0,0,0)$. ■

In Example 4.2, a proper Markov chain and waiting time random variable have been defined to compute the reliability of the system. Such a method is closely related to the Markov chain imbedding approach to study the reliability of systems. See, e.g. (Koutras 1996).

As illustrated by Examples 4.1 and 4.2, phase-type distributions are very efficient to derive exact distribution of a waiting time random variable which can be represented as the time until absorption in a Markov chain. In the sequel, we present useful properties of discrete phase-type distributions.

The probability generating function given by (4.3) is rational and it can be represented as

$$\phi(z) = \pi_0 + \frac{a_1 z + \ldots + a_m z^m}{1 + b_1 z + \ldots + b_m z^m}. \tag{4.5}$$

By using the coefficients in (4.5), the probability mass function of T can be recursively computed from:

$$P\{T=1\}=a_1,$$
$$P\{T=n\}=a_n-\sum_{j=1}^{\min(m,n-1)} b_jP\{T=n-j\},\ \ n>1, \tag{4.6}$$

where $a_n=0$ for $n>m$ (see, e.g. Eisele 2006). That is, if the probability generating function of a phase-type random variable T is known, then the probability mass function of T can be calculated recursively from (4.6). This is useful since in some cases the derivation of the probability generating function is simpler than obtaining directly the probability distribution. For the negative Binomial distribution with probability mass function

$$P\{T=t\}=\binom{t-1}{r-1}p^r(1-p)^{t-r}\ ,t=r,r+1,...,$$

we have

$$a_i=\begin{cases} p^r, & \text{if } i=r \\ 0 & \text{otherwise}\end{cases}$$

and

$$b_i=(-1)^i\binom{r}{i}(1-p)^i,$$

for $i=1,2,...r$. Thus, we can obtain the following recursive equation to compute $P\{T=t\}$:

$$P\{T=r\}=p^r,$$
$$P\{T=t\}=-\sum_{j=1}^{r}(-1)^j\binom{r}{j}(1-p)^jP\{T=t-j\},\ \ t>r,$$
$$P\{T=t\}=0,\ t<r.$$

For a phase-type random variable having probability generating function (4.5), assume that the characteristic polynomial

$$\lambda^m+b_1\lambda^{m-1}+...+b_{m-1}\lambda+b_m=0$$

has real distinct roots $\lambda_0,\lambda_1,...,\lambda_{m-1}$. For such a phase-type random variable, (Eryilmaz 2021) obtained the following mixture representation:

$$P\{T=t\}=\sum_{k=0}^{m-1}\omega_k\lambda_k^{t-1}(1-\lambda_k),$$

where

$$\omega_k=\frac{1}{(1-\lambda_k)\prod\limits_{\substack{i=0\\ i\neq k}}^{m-1}(\lambda_k-\lambda_i)}\sum_{j=0}^{m-1}(-1)^jp(m-j)\sum_{\substack{0\le i_1<...<i_j\le m-1\\ i_j\neq k}}\cdots\sum\lambda_{i_1}...\lambda_{i_j},$$

for $k = 0, 1, ..., m-1$, where $p(t) = P\{T = t\}$. The real coefficients $\omega_0, \omega_1, ..., \omega_{m-1}$ satisfy

$$\sum_{k=0}^{m-1} \omega_k = 1.$$

(Kan and Eryilmaz 2021) obtained an equation to compute the kth moment of T recursively using the coefficients given in (4.5). If T has a phase-type distribution with probability generating function (4.5), then

$$E(T^k) = \frac{\sum_{n=1}^{m} a_n n^k - \sum_{n=1}^{m} b_n \sum_{i=0}^{k-1} \binom{k}{i} n^{k-i} E(T^i)}{1 + \sum_{n=1}^{m} b_n}.$$

In particular, the first and second moments of T can be computed from

$$E(T) = \frac{\sum_{n=1}^{m} n(a_n - b_n)}{1 + \sum_{n=1}^{m} b_n},$$

and

$$E(T^2) = \frac{\sum_{n=1}^{m} a_n n^2 - \sum_{n=1}^{m} b_n \left[n^2 + nE(T)\right]}{1 + \sum_{n=1}^{m} b_n}.$$

Example 4.3. Recall the probability generating function of geometric distribution of order k that has been previously presented in section 2.

$$E(z^{T_k}) = \frac{p^k z^k - p^{k+1} z^{k+1}}{1 - z + (1-p) p^k z^{k+1}}. \tag{4.7}$$

Because $E(z^{T_k})$ has the same form with (4.5), the geometric distribution of order k can be represented as a phase-type distribution. Indeed, (4.7) is same with (4.5) when $\pi_0 = 0$, $m = k+1$ and

$$a_1 = 0, ..., a_{k-1} = 0, a_k = p^k, a_{k+1} = -p^{k+1},$$
$$b_1 = -1, b_2 = 0, ..., b_k = 0, b_{k+1} = (1-p)p^k.$$

Thus, from (4.6),

$$P\{T_k = n\} = P\{T_k = n-1\} - (1-p)p^k P\{T_k = n-k-1\}, \tag{4.8}$$

for $n > k+1$. The Equation (4.8) agrees with the Equation (2.8). ■

Because geometric distribution of order k is phase-type, matrix-based equation can be easily obtained by defining an appropriate Markov chain. This is presented in the following example.

Example 4.4. Geometric distribution of order k is the distribution of the number of trials until k consecutive successes in a sequence of Bernoulli trials. To find a phase representation for the geometric distribution of order k, let us define a Markov chain on the states $\{0,1,...,k\}$, where the absorbing state $\{0\}$ represents k consecutive successes $\underbrace{11...1}_{k}$. The transient state $\{i\}$ is defined by $\underbrace{11...1}_{i-1}$ when $i=2,...,k$ and the state $\{1\}$ corresponds to the failure. Then, $T_k \sim PH_k(\pi,\mathbf{Q})$, where $\pi=(1,0,...,0)$, and

$$\mathbf{Q}=\begin{bmatrix} 1-p & p & 0 & \dots & 0 \\ 1-p & 0 & p & \dots & 0 \\ \vdots & \vdots & \vdots & & \vdots \\ 1-p & 0 & 0 & \dots & 0 \end{bmatrix}_{k\times k},$$

where p is the probability of success for a Bernoulli trial (Tank and Eryilmaz 2015). ■

Example 4.5. Successive tests are applied to a certain equipment before purchasing it. Such a test procedure is known to be start-up demonstration test. Each start-up can result either as a failure or success, and a decision about the acceptance/rejection of the equipment is made depending one some rules corresponding to failures/successes. According to consecutive successes total failures (CSTF) test, the equipment is accepted if k consecutive successes are observed before d failures (Balakrishnan and Chan 2000). Thus, the test length, i.e. the number of start-ups until acceptance/rejection of the equipment is defined by

$$T_{k,d}=\min(T_k,W_d),$$

where T_k is the number of start-ups until k consecutive successful start-ups and W_d is the number of start-ups until d failures. In the following, we obtain the distribution of $T_{k,d}$ when $k=4$ and $d=2$ using phase-type modeling. Consider a Markov chain with two components such that the first component counts the cumulative number of failures (unsuccessful start-ups) and the second component records the number of consecutive successful start-ups. Thus, the transient states for the Markov chain under concern are $(0,0),(0,1),(0,2),(0,3),(1,0),(1,1),(1,2),(1,3)$. The test ends either with $d=2$ total failures or $k=4$ consecutive successes. Therefore, the absorbing states are $(2,0),(0,4),(1,4)$. Let p be the probability that individual start-up results as a success. Then, $T_{4,2}\sim PH_8(\pi,\mathbf{Q})$, where $\pi=(1,0,0,0,0,0,0,0)$, and

$$\mathbf{Q}=\begin{bmatrix} 0 & p & 0 & 0 & 1-p & 0 & 0 & 0 \\ 0 & 0 & p & 0 & 1-p & 0 & 0 & 0 \\ 0 & 0 & 0 & p & 1-p & 0 & 0 & 0 \\ 0 & 0 & 0 & 0 & 1-p & 0 & 0 & 0 \\ 0 & 0 & 0 & 0 & 0 & p & 0 & 0 \\ 0 & 0 & 0 & 0 & 0 & 0 & p & 0 \\ 0 & 0 & 0 & 0 & 0 & 0 & 0 & p \\ 0 & 0 & 0 & 0 & 0 & 0 & 0 & 0 \end{bmatrix}.$$

Using

$$E(T_{4.2}) = \pi(\mathbf{I}-\mathbf{Q})^{-1}\mathbf{e}',$$

the mean test length is found to be

$$E(T_{4.2}) = \frac{1-p^4}{(1-p)p^4}\left[1-(1-p^4)^2\right].$$

If for example $p = 0.8$, then the test is terminated on or before the 5th trial is

$$P\{T_{4.2} \leq 5\} = 1 - \pi\mathbf{Q}^5\mathbf{e}' = 0.7542. \quad \blacksquare$$

Example 4.6. (Zhao et al. 2018) defined and studied a mixed shock model with a change point. The model is called a two-stage failure shock model and combines cumulative and run shock models with a change point. In particular, the change point appears when a run of damaging shocks with a fixed length k_c occurs. Stage 1 is the process from new to the change point. In stage 2, the system fails if either the number of cumulative damaging shocks reachs a critical value $(k_{tf} - k_c)$ or the number of consecutive damaging shocks is above a fixed value k_{cf} . (Zhao et al. 2018) obtained phase representation for the total number of shocks until the change point and the total number of shocks until the system failure by defining a trivariate Markov chain $X_n = (u_n, v_n, w_n)$, where

$$u_n = \begin{cases} 0, & \text{if the system is in stage 1} \\ 1, & \text{if the system is in stage 2} \\ 2, & \text{if the system is failed,} \end{cases}$$

v_n denotes the number of last run of damaging shocks in stage 2, and w_n is the number of last run of damaging shocks in stage 1 when $u_n = 0$; denotes the number of cumulative damaging shocks in stage 2 when $u_n = 1$. Let $k_c = 2, k_{tf} = 3$, and $k_{cf} = 2$. Then, the corresponding Markov chain has three transient states, namely, $(0,0,0), (0,0,1)$, and $(1,0,2)$. If N_c and N_f denote respectively the number of shocks until the change point and the number of shocks until the system failure, then from the results of (Zhao et al. 2018), we can obtain $N_c \sim PH_2(\beta, \mathbf{A})$ and $N_f \sim PH_3(\pi, \mathbf{Q})$, where $\beta = (1,0), \pi = (1,0,0)$, and

$$A = \begin{bmatrix} q & p \\ q & 0 \end{bmatrix}, \quad \mathbf{Q} = \begin{bmatrix} q & p & 0 \\ q & 0 & p \\ 0 & 0 & q \end{bmatrix},$$

where p is the probability of a damaging shock and q is the probability of non-damaging shock. $\blacksquare$

It should be noted that any discrete distribution having a finite support is a discrete phase-type distribution. Consider the Binomial distribution with probability mass function

$$P\{T = t\} = \binom{n}{t} p^t (1-p)^{n-t},$$

for $t = 0, 1, ..., n$. Then, we have

$$P\{T = 0\} = 1 - \pi \mathbf{e}',$$

and for $t = 1, 2, ..., n$,

$$P\{T = t\} = \pi \mathbf{Q}^{t-1} \mathbf{u}',$$

where $\pi = (\pi_1, ..., \pi_n)$ with

$$\pi_i = \binom{n}{i} p^i (1-p)^{n-i},$$

and

$$\mathbf{Q} = \begin{bmatrix} 0 & 0 & \dots & 0 & 0 \\ 1 & 0 & \dots & 0 & 0 \\ 0 & 1 & \dots & 0 & 0 \\ \vdots & \vdots & \ddots & & \vdots \\ 0 & 0 & & 1 & 0 \end{bmatrix}_{n \times n}.$$

4.2 CLOSURE AND OTHER PROPERTIES

Phase-type distributions have very useful closure properties. For example, the sum, minimum, and maximum of independent phase-type random variables have phase-type distributions. In the following, we present some properties of discrete phase-type distributions along with illustrative examples. The proofs of the first four results can be found in (Neuts 1975).

Let T_1 and T_2 be two independent random variables such that $T_1 \sim PH_m(\pi, \mathbf{Q}_1)$ and $T_2 \sim PH_n(\sigma, \mathbf{Q}_2)$. Note that the dimensions of $\mathbf{e}'$ and $\mathbf{I}$ are appropriately chosen in all equations.

Property 1. $T_1 + T_2 \sim PH_{m+n}(\gamma, \mathbf{L})$, where

$$\gamma = (\pi_1, ..., \pi_m, \pi_{m+1}\sigma_1, \pi_{m+1}\sigma_2, ..., \pi_{m+1}\sigma_n),$$

and $\mathbf{L}$ is an $m+n$ by $m+n$ matrix which has the form

$$\mathbf{L} = \begin{bmatrix} \mathbf{Q}_1 & (\mathbf{I} - \mathbf{Q}_1)\mathbf{e}'\sigma \\ \mathbf{0} & \mathbf{Q}_2 \end{bmatrix}.$$

Example 4.7. A system consists of one active and one standby nonrepairable units. The system starts its operation with the active unit and the standby unit is put into operation upon the failure of the active unit. The standby unit does not fail while it is inactive. Such a standby unit is called cold standby. The lifetimes of the units are both integer valued. Let T_1 and T_2 denote respectively the lifetimes of active and standby units. Assume that

$$P\{T_1 = t\} = p_1(1 - p_1)^{t-1},$$

and

$$P\{T_2 = t\} = p_2(1-p_2)^{t-1}, t = 1,2,\ldots.$$

Clearly, the lifetime of the entire system is $T_1 + T_2$, and $T_1 \sim PH_1(1, 1-p_1)$ and $T_2 \sim PH_1(1, 1-p_2)$. Thus,

$$T_1 + T_2 \sim PH_2(\gamma, \mathbf{L}),$$

where $\gamma = (1,0)$ and

$$\mathbf{L} = \begin{bmatrix} 1-p_1 & p_1 \\ 0 & 1-p_2 \end{bmatrix}.$$

Thus, the probability that the system survives at a time t can be computed from

$$P\{T_1 + T_2 > t\} = (1,0) \begin{bmatrix} 1-p_1 & p_1 \\ 0 & 1-p_2 \end{bmatrix}^t \begin{bmatrix} 1 \\ 1 \end{bmatrix},$$

$t = 1,2,\ldots.$

Property 2. $\min(T_1, T_2) \sim PH_{m\times n}(\gamma, \mathbf{L})$, where

$$\gamma = \pi \otimes \sigma,$$

and

$$\mathbf{L} = \mathbf{Q}_1 \otimes \mathbf{Q}_2,$$

where $\otimes$ denotes Kronecker product. For two matrices $\mathbf{Q}_1 = (q_{ij}), 1 \le i,j \le m$ with dimension $m \times m$ and $\mathbf{Q}_2$ with dimension $n \times n$,

$$\mathbf{Q}_1 \otimes \mathbf{Q}_2 = \begin{bmatrix} q_{11}\mathbf{Q}_2 & q_{12}\mathbf{Q}_2 & \cdots & q_{1m}\mathbf{Q}_2 \\ q_{21}\mathbf{Q}_2 & q_{22}\mathbf{Q}_2 & \cdots & q_{2m}\mathbf{Q}_2 \\ \vdots & \vdots & \ddots & \vdots \\ q_{m1}\mathbf{Q}_2 & q_{m2}\mathbf{Q}_2 & \cdots & q_{mm}\mathbf{Q}_2 \end{bmatrix}.$$

Property 3. $\max(T_1, T_2) \sim PH_{mn+m+n}(\gamma, \mathbf{L})$, where

$$\gamma = (\pi \otimes \sigma, \pi\sigma_{n+1}, \pi_{m+1}\sigma),$$

and

$$\mathbf{L} = \begin{bmatrix} \mathbf{Q}_1 \otimes \mathbf{Q}_2 & \mathbf{Q}_1 \otimes ((\mathbf{I}-\mathbf{Q}_2)\mathbf{e}') & ((\mathbf{I}-\mathbf{Q}_1)\mathbf{e}') \otimes \mathbf{Q}_2 \\ \mathbf{0} & \mathbf{Q}_1 & \mathbf{0} \\ \mathbf{0} & \mathbf{0} & \mathbf{Q}_2 \end{bmatrix}.$$

Example 4.8. Consider a system consisting of three independent components. Let T_i denote the lifetime of the ith component, $i = 1,2,3$. Assume that the subsystem 1 has the component with lifetime T_1 and subsystem 2 consists of the components with lifetimes T_2 and T_3. The components with lifetimes T_2 and T_3 have a parallel connection in subsystem 2. The two subsystems are connected in series. Thus, the lifetime of the entire system is

$$T = \min(T_1, \max(T_2, T_3)).$$

Assume that

$$P\{T_1 = t\} = p_1(1-p_1)^{t-1},$$

and

$$P\{T_i = t\} = p_2(1-p_2)^{t-1}, i = 2,3; t = 1,2,....$$

The problem is to compute the mean time to failure of the system, i.e. $E(T)$ as a function of the parameters p_1 and p_2. From Property 3, we have $\max(T_2,T_3) \sim PH_3(\beta,\mathbf{M})$, where

$$\beta = (1,0,0),$$

and

$$\mathbf{M} = \begin{bmatrix} (1-p_2)^2 & (1-p_2)p_2 & p_2(1-p_2) \\ 0 & 1-p_2 & 0 \\ 0 & 0 & 1-p_2 \end{bmatrix}.$$

Now, using Property 2, we obtain $T \sim PH_3(\gamma,\mathbf{L})$, where

$$\gamma = (1,0,0),$$

and

$$\mathbf{L} = \begin{bmatrix} (1-p_1)(1-p_2)^2 & (1-p_1)(1-p_2)p_2 & (1-p_1)(1-p_2)p_2 \\ 0 & (1-p_1)(1-p_2) & 0 \\ 0 & 0 & (1-p_1)(1-p_2) \end{bmatrix}.$$

Thus, the mean time to failure of the system can be computed from

$$\begin{aligned} E(T) &= \gamma(\mathbf{I}-\mathbf{L})^{-1}\mathbf{e}' \\ &= \frac{p_1 + 3p_2 + 2p_1p_2^2 - 2p_2^2 - 3p_1p_2}{(p_1+p_2-p_1p_2)(p_1+2p_2+p_1p_2^2-p_2^2-2p_1p_2)}. \end{aligned}$$

We compute $E(T)$ for selected values of p_1 and p_2 :

$$\begin{aligned} p_1 = 0.1\ ,\ p_2 = 0.2 &\Longrightarrow E(T) = 4.7844, \\ p_1 = 0.1\ ,\ p_2 = 0.5 &\Longrightarrow E(T) = 2.3460, \\ p_1 = 0.25\ ,\ p_2 = 0.2 &\Longrightarrow E(T) = 3.0769. \end{aligned}$$

Property 4. Let N be an integer valued random variable having phase-type distribution and $Y_1, Y_2, ...$ be a sequence of integer valued, and independent and identically distributed random variables. If $N \sim PH_m(\pi,\mathbf{Q}_1)$ and $Y_i \sim PH_n(\sigma,\mathbf{Q}_2), i = 1,2,...,$ and N and $Y_1, Y_2, ...$ are independent, then

$$\sum_{i=1}^{N} Y_i \sim PH_{m\times n}(\gamma,\mathbf{L}),$$

where

$$\gamma = \sigma \otimes \pi\,(\mathbf{I} - \alpha\mathbf{Q}_1)^{-1},$$

and

$$\mathbf{L} = \mathbf{Q}_2 \otimes \mathbf{I} + ((\mathbf{I} - \mathbf{Q}_2)\mathbf{e}')\sigma \otimes (\mathbf{I} - \alpha \mathbf{Q}_1)^{-1} \mathbf{Q}_1,$$

with $\alpha = 1 - \sigma \mathbf{e}'$.

In Property 4, both N and Y_is are assumed to have phase-type distributions. Eisele (2006) obtained a recursive equation to compute the probability mass function of the compound random variable $S = \sum_{i=1}^{N} Y_i$ when $N \sim PH_m(\pi, \mathbf{Q}_1)$ and $Y_1, Y_2, \ldots$ is a sequence of positive valued independent and identically distributed random variables with common probability mass function $f_Y(y) = P\{Y_i = y\}$, independent of N. The corresponding equation is given by

$$P\{S = t\} = \sum_{j=1}^{\min(m,t)} a_j f_Y^{*j}(t) - \sum_{j=1}^{\min(m,t-1)} b_j \left(\sum_{u=1}^{t-1} P\{S = u\} f_Y^{*j}(t-u) \right), \tag{4.9}$$

for $t \geq 1$, where f_Y^{*j} is the jth convolution of f_Y, and $b_1, \ldots, b_d$ are coefficients of the characteristic polynomial of $\mathbf{Q}_1$, namely

$$\det(z\mathbf{I} - \mathbf{Q}_1) = z^m + \sum_{i=1}^{m} b_i z^{m-i}, \tag{4.10}$$

and the coefficients $a_1, \ldots, a_d$ are computed through the following equations:

$$a_n = P\{N = n\} + \sum_{i=1}^{n-1} b_i P\{N = n - i\}, \tag{4.11}$$

for $n = 2, \ldots, d$ with $a_1 = P\{N = 1\}$.

Example 4.9. For a system that is subject to shocks over time, let $Y_1, Y_2, \ldots$ be the times between arrivals of shocks. The magnitude of the ith shock is denoted by Z_i, $i \geq 1$. The system fails if the magnitudes of k consecutive shocks exceed or equal to a prespecified level $c > 0$. This is the run shock model studied by (Mallor and Omey 2001). The lifetime of the system is then defined by the random variable $S = \sum_{i=1}^{T_k} Y_i$, where T_k has geometric distribution of order k with $p = P\{Z \geq c\}$. Assume that

$$f(y) = P\{Y_i = y\} = \theta(1-\theta)^{y-1}, y = 1, 2, \ldots.$$

In Example 4.4, it has been shown that $T_k \sim PH_k(\pi, \mathbf{Q})$, where $\pi = (1, 0, \ldots, 0)$, and

$$\mathbf{Q} = \begin{bmatrix} 1-p & p & 0 & \ldots & 0 \\ 1-p & 0 & p & \ldots & 0 \\ \vdots & \vdots & \vdots & & \vdots \\ 1-p & 0 & 0 & \ldots & 0 \end{bmatrix}.$$

Because $Y_i \sim PH_1(1, 1-\theta)$, using Property 4, the lifetime of the system has phase-type distribution with representation $S \sim PH_k(\gamma, \mathbf{L})$, where $\gamma = (1, 0, \ldots, 0)$, and the $k \times k$ matrix $\mathbf{L}$ is given by

$$\mathbf{L} = \begin{bmatrix} 1-\theta+\theta(1-p) & \theta p & 0 & 0 & \dots & 0 & 0 \\ \theta(1-p) & 1-\theta & \theta p & 0 & \dots & 0 & 0 \\ \vdots & \vdots & \vdots & \ddots & & \vdots & \vdots \\ \theta(1-p) & 0 & 0 & 0 & & 1-\theta & \theta p \\ \theta(1-p) & 0 & 0 & 0 & \dots & 0 & 1-\theta \end{bmatrix}$$

(Koutras and Eryilmaz 2017).

On the other hand, for the random variable $T_k \sim PH_k(\pi, \mathbf{Q})$, using (4.10) and (4.11), we can obtain $b_i = p^i - p^{i-1}, i = 1, ..., k$ and

$$a_i = \begin{cases} p^k, & \text{if } i = k \\ 0, & \text{otherwise} \end{cases}$$

Thus, a recursive equation for the distribution of S is immediately obtained from (4.9). For $t > k$,

$$\begin{aligned} P\{S=t\} &= \binom{t-1}{k-1} (\theta p)^k (1-\theta)^{t-k} \\ &\quad - \sum_{j=1}^{k} (p^j - p^{j-1}) \left(\sum_{u=1}^{t-j} P\{S=u\} \binom{t-u-1}{j-1} \theta^j (1-\theta)^{t-u-j} \right) \end{aligned}$$

with initial conditions $P\{S=k\} = (\theta p)^k$ and $P\{S=t\} = 0$ if $t < k$ (Koutras and Eryilmaz 2017). It should be noted that when $k = 1$, the distribution of S is same with the geometric distribution with probability mass function

$$P\{S=t\} = (\theta p)(1-\theta p)^{t-1}, t = 1, 2, \dots.$$

In Table 4.2, we compute the survival probability $P\{S > t\}$ for selected values of the parameters for $p = 0.2$ when $k = 2, 3$. As expected, an increase in k leads to an increase in survival probability.

Table 4.2
Survival Probability of the System When $p = 0.2$

k	θ	t	$P\{S>t\}$	k	θ	t	$P\{S>t\}$
2	0.1	5	0.9965	3	0.1	5	0.9999
		10	0.9867			10	0.9993
		15	0.9738			15	0.9981
2	0.5	5	0.9445	3	0.5	5	0.9946
		10	0.8666			10	0.9792
		15	0.7948			15	0.9634

Example 4.10. For a system that is subject to shocks over time, let $Y_1, Y_2, \dots$ denote the times between arrivals of shocks. The magnitude of the ith shock is denoted by Z_i, $i \geq 1$. The system fails upon the occurrence of k consecutive shocks of size at least d_1 or a single large shock of size at least d_2, $d_1 < d_2$. This is the mixed shock

model studied by (Eryilmaz and Tekin 2019). If T denotes the number of shocks until k consecutive shocks above or equal to d_1 or a single shock of size at least d_2. Then, the lifetime of the system is $S = \sum_{i=1}^{T} Y_i$. (Eryilmaz and Tekin 2019) obtained phase representation for T by defining a Markov chain that is based on a sequence of trinary trials $I_1, I_2, ...$ such that

$$I_t = \begin{cases} 0, & \text{if } Z_t < d_1 \\ 1, & \text{if } d_1 \leq Z_t < d_2 \\ 2, & \text{if } Z_t \geq d_2. \end{cases}$$

Let $k = 2$. In this case, the system fails if either there are two consecutive shocks of size at least d_1 or a single shock of size at least d_2. Then, the associated Markov chain has five states which correspond to $0, 1, 2, 11$, and 12. The absorbing states correspond to $2, 11$, and 12. If $p_1 = P\{d_1 \leq Z_t < d_2\}$ and $p_2 = P\{Z_t \geq d_2\}$, $t = 1, 2, ...$, then $T \sim PH_2(\pi, \mathbf{Q})$ with $\pi = (1, 0)$, and

$$\mathbf{Q} = \begin{bmatrix} 1 - p_1 - p_2 & p_1 \\ 1 - p_1 - p_2 & 0 \end{bmatrix}.$$

Our aim is to find the mean time to failure of the system. It can be easily shown that

$$\begin{aligned} E(T) &= \pi(\mathbf{I} - \mathbf{Q})^{-1}\mathbf{e}' \\ &= \frac{1 - p_1^2}{(1 - p_1)p_1^2 + p_2(1 - p_1^2)}. \end{aligned}$$

Thus, the mean time to failure of the system is found to be

$$E(S) = \mu \left[\frac{1 - p_1^2}{(1 - p_1)p_1^2 + p_2(1 - p_1^2)} \right],$$

where $\mu = E(Y)$. Let $p_1 = 0.2$, $p_2 = 0.1$, and $\mu = 1$. Then,

$$E(S) = 7.5.$$

Property 5. Let $T \sim PH_m(\pi, \mathbf{Q})$. Then,

$$(T - t \mid T > t) \sim PH_m\left(\frac{\pi\mathbf{Q}^t}{\pi\mathbf{Q}^t\mathbf{e}'}, \mathbf{Q}\right),$$

and

$$E(T - t \mid T > t) = \frac{\pi\mathbf{Q}^t}{\pi\mathbf{Q}^t\mathbf{e}'}(\mathbf{I} - \mathbf{Q})^{-1}\mathbf{e}'. \tag{4.12}$$

Example 4.11. Let T denote the lifetime of a certain bearing whose lifetime is measured in terms of the number of its rotations. Assume that the distribution of T is a mixture of two geometric distributions with positive weights, i.e. $T \sim PH_2(\pi, \mathbf{Q})$, where $\pi = (\omega_1, \omega_2)$, and

$$\mathbf{Q} = \begin{bmatrix} 1 - p_1 & 0 \\ 0 & 1 - p_2 \end{bmatrix},$$

with $\omega_i > 0, i = 1,2$, and $\omega_1 + \omega_2 = 1$. Suppose that $\omega_1 = 0.25, \omega_2 = 0.75$, $p_1 = 0.0003$, and $p_2 = 0.0002$. Let us find the mean residual lifetime of the bearing given that it has already completed 3000 rotations. Using (4.12), the desired value is found to be

$$E(T - 3000 \mid T > 3000) \simeq 4670 \text{ rotations.} \quad \blacksquare$$

The proof of the following property can be found in Eryilmaz (2019).

Property 6. Let $T_1 \sim PH_m(\pi, \mathbf{Q}_1)$ and $T_2 \sim PH_n(\sigma, \mathbf{Q}_2)$. Then,

$$P\{T_1 < T_2\} = 1 - (\pi \otimes \sigma)\left(\mathbf{I} - (\mathbf{Q}_1 \otimes \mathbf{Q}_2)\right)^{-1}\left(\mathbf{e}' \otimes \mathbf{v}'\right), \tag{4.13}$$

where $\mathbf{v}' = (\mathbf{I} - \mathbf{Q}_2)\mathbf{e}'$.

Example 4.12. Consider a system consisting of two identical units. At time $t = 0$, both of the two units are new, and one of the units starts working while the other unit is in a cold standby state, i.e. it does not fail when it is inactive. The standby unit is put into operation upon the failure of the active component and the failed component is executed repair. That is, the standby unit becomes active while the failed unit is under repair. Let T_1 denote the lifetime of the active unit and T_2 be the time for repair of the failed unit. Let us find the probability that the active unit fails before the completion of the repair of the other unit when $P\{T_1 = t\} = p_1(1-p_1)^{t-1}, t = 1,2,...$ and $T_2 \sim PH_2(\sigma, \mathbf{Q}_2)$ with $\sigma = (1,0)$ and

$$\mathbf{Q}_2 = \begin{bmatrix} 1-p_2 & p_2 \\ 0 & 1-p_2 \end{bmatrix}.$$

Because $T_1 \sim PH_1(1, 1-p_1)$, using (4.13),

$$P\{T_1 < T_2\} = \frac{p_1(p_1 + 2p_2 + p_1 p_2^2 - p_2^2 - 2p_1 p_2)}{(p_1 + p_2 - p_1 p_2)^2}.$$

If for example $p_1 = 0.1$ and $p_2 = 0.5$, then the probability that the active unit fails before the completion of the repair of the other unit is found to be

$$P\{T_1 < T_2\} = 0.2562. \quad \blacksquare$$

Example 4.13. Consider a series system with two independent components denoted by 1 and 2. Assume that the components are equipped with cold standby components denoted by $1'$ and $2'$. Let T_1 and T_2 denote the lifetimes of the active units, and T_1^1 and T_2' represent the lifetimes of the standby units. Then, the lifetime of the entire system is

$$T = \min(T_1 + T_1', T_2 + T_2').$$

Assume that

$$P\{T_1 = t\} = p_1(1-p_1)^{t-1}, \quad P\{T_1' = t\} = p_1(1-p_1)^{t-1},$$
$$P\{T_2 = t\} = p_2(1-p_2)^{t-1}, \quad P\{T_2' = t\} = p_2(1-p_2)^{t-1},$$

$t = 1, 2, \ldots$. Clearly, the lifetime of the entire system is $T_1 + T_2$, and $T_1 \sim PH_1(1, 1 - p_1)$ and $T_2 \sim PH_1(1, 1 - p_2)$. Thus,

$$T_1 + T_1' \sim PH_2((1,0), \begin{bmatrix} 1-p_1 & p_1 \\ 0 & 1-p_1 \end{bmatrix}),$$

and

$$T_2 + T_2' \sim PH_2((1,0), \begin{bmatrix} 1-p_2 & p_2 \\ 0 & 1-p_2 \end{bmatrix}).$$

Thus, using Property 2,

$$\min(T_1 + T_1', T_2 + T_2') \sim PH_4(\gamma, \mathbf{L}),$$

where

$$\gamma = (1,0) \otimes (1,0) = (1,0,0,0)$$

and

$$\mathbf{L} = \begin{bmatrix} 1-p_1 & p_1 \\ 0 & 1-p_1 \end{bmatrix} \otimes \begin{bmatrix} 1-p_2 & p_2 \\ 0 & 1-p_2 \end{bmatrix}$$
$$= \begin{bmatrix} (1-p_1)(1-p_2) & (1-p_1)p_2 & p_1(1-p_2) & p_1p_2 \\ 0 & (1-p_1)(1-p_2) & 0 & p_1(1-p_2) \\ 0 & 0 & (1-p_1)(1-p_2) & (1-p_1)p_2 \\ 0 & 0 & 0 & (1-p_1)(1-p_2) \end{bmatrix}.$$

Thus, the system's survival function can be computed from

$$P\{T > t\} = \gamma \mathbf{L}^t \mathbf{e}',$$

$t = 1, 2, \ldots$. ■

In the following example, we consider a discrete time queueuing system when the times between arrivals follow geometric distribution and the service time distribution has a general discrete phase-type distribution. This is the generalization of the model previously considered in Example 3.3. The detailed analysis of such discrete time queues and their generalizations have been presented by Alfa (2016).

Example 4.14. Recall from Example 3.3 that the failed machines arrive at the maintenance center for repair at time marks. Assume that no more than one machine may arrive at a given time mark and the arrivals occur independently. The distribution of the interarrival time between successive arrivals follows geometric distribution with mean $\frac{1}{p}$. The failed machines are served in their order of arrival and the service time distribution is phase type with representation $PH_m(\beta, \mathbf{S})$. Let X_n denote the number of customers (machines) in the system at time n. Then, for $\pi_0 = \lim_{n \longrightarrow \infty} P\{X_n = 0\}$ and $\pi_i = [\pi_{i.1}, \pi_{i.2}, \ldots, \pi_{i.m}]$, the probability distribution $(\pi_0, \pi_1, \pi_2, \ldots)$ can be computed from

$$\pi_{i+1} = \pi_i \mathbf{R},$$

for $i \geq 1$, with

$$\pi_0 = \pi_0(1-p) + \pi_1\mathbf{E},\ \pi_1 = \pi_0\mathbf{C} + \pi_1(\mathbf{A}_1 + \mathbf{R}\mathbf{A}_2),$$

and $\pi_0 + \pi_1(\mathbf{I}-\mathbf{R})^{-1}\mathbf{e}' = 1$, where

$$\mathbf{R} = \mathbf{A}_0\left[\mathbf{I} - \mathbf{A}_1 - \mathbf{A}_0\mathbf{e}'\beta\right]^{-1},$$

and

$$\begin{array}{c} \mathbf{A}_0 = p\mathbf{S}\ ,\ \mathbf{A}_1 = p(\mathbf{I}-\mathbf{S})\mathbf{e}'\beta + (1-p)\mathbf{S} \\ \mathbf{C} = p\beta\ ,\ \mathbf{E} = (1-p)(\mathbf{I}-\mathbf{S})\mathbf{e}' \end{array}$$

(Alfa 2016). For an illustration, let the service time distribution have phase type distribution of order $m = 2$ with $\beta = (1,0)$, and

$$\mathbf{S} = \begin{bmatrix} 1-d & d \\ 1-d & 0 \end{bmatrix}.$$

Then,

$$\mathbf{R} = \begin{bmatrix} \frac{p(1-d^2(1-p))}{d^2(1-p)^2} & \frac{p}{d(1-p)} \\ \frac{p(1-d)}{d^2(1-p)^2} & \frac{p(1-d)}{d(1-p)} \end{bmatrix},$$

and

$$\pi_0 = 1 - \frac{p(d+1)}{d^2},$$

$$\pi_1 = \left[\frac{p}{d^2(1-p)^2}\pi_0, \frac{p}{d(1-p)}\pi_0\right],$$

and $\pi_{i+1} = \pi_i\mathbf{R}$, for $i \geq 1$. The mean number of the customers in the system can now be computed from

$$\pi_1(\mathbf{I}-\mathbf{R})^{-2}\mathbf{e}' = \frac{p(1-2p-d(p-1))}{d^2 - p(d+1)},$$

for $\frac{p(d+1)}{d^2} < 1$. ■

Example 4.15. Let the random variable T_k have geometric distribution of order k which has been considered in Example 4.4. Then, $T_k \sim PH_k(\pi, \mathbf{Q})$, where $\pi = (1,0,\ldots,0)$, and

$$\mathbf{Q} = \begin{bmatrix} 1-p & p & 0 & \ldots & 0 \\ 1-p & 0 & p & \ldots & 0 \\ \vdots & \vdots & \vdots & & \vdots \\ 1-p & 0 & 0 & \ldots & 0 \end{bmatrix}_{k \times k}.$$

The conditional random variable $(T_k - t \mid T_k > t)$ represents the residual time (or equivalently the number of trials) until k consecutive successes given that no success run of length k occurred by time t. From (4.12) we have

$$m_k(t) = E(T_k - t \mid T_k > t) = \frac{\pi\mathbf{Q}^t}{\pi\mathbf{Q}^t\mathbf{e}'}(\mathbf{I}-\mathbf{Q})^{-1}\mathbf{e}'.$$

Below, we compute the mean residual time until k consecutive successes for selected values of k and p:

$$p = 0.5 \Longrightarrow m_3(5) = 12.4167, m_4(5) = 27.6552$$
$$p = 0.65 \Longrightarrow m_3(5) = 5.9872, m_4(5) = 10.9710$$
$$p = 0.95 \Longrightarrow m_3(5) = 1.4877, m_4(5) = 2.5507.$$

As expected, $m_k(t)$ is decreasing in p and increasing in k.

Example 4.16. Consider the system described in Example 4.8. The lifetime of the system is

$$T = \min(T_1, \max(T_2, T_3)).$$

It has been shown that $T \sim PH_3(\gamma, \mathbf{L})$, where

$$\gamma = (1,0,0),$$

and

$$\mathbf{L} = \begin{bmatrix} (1-p_1)(1-p_2)^2 & (1-p_1)(1-p_2)p_2 & (1-p_1)(1-p_2)p_2 \\ 0 & (1-p_1)(1-p_2) & 0 \\ 0 & 0 & (1-p_1)(1-p_2) \end{bmatrix}.$$

The mean residual lifetime of the system can be computed from

$$E(T-t \mid T>t) = \frac{\gamma \mathbf{L}^t}{\gamma \mathbf{L}^t \mathbf{e}'} (\mathbf{I}-\mathbf{L})^{-1} \mathbf{e}'.$$

Let $p_1 = 0.1$ and $p_2 = 0.2$. Then,

$$E(T-10 \mid T>10) = 3.6402,$$
$$E(T-15 \mid T>15) = 3.5931,$$
$$E(T-20 \mid T>20) = 3.5785,$$
$$E(T-50 \mid T>50) = 3.5714. \quad \blacksquare$$

Consider a discrete renewal process when the times between events follow phase-type distribution. From the equation (2.24), the renewal function is

$$M(n) = F(n) + \sum_{t=1}^{n-1} M(n-t) p_t.$$

Using (4.1) and (4.2), the renewal function becomes

$$M(n) = 1 - \pi \mathbf{Q}^n \mathbf{e}' + \sum_{t=1}^{n-1} M(n-t) \pi \mathbf{Q}^{t-1} \mathbf{u}'.$$

Let the times between events follow geometric distribution of order $k = 2$, i.e.

$$\begin{aligned} p_t &= \pi \mathbf{Q}^{t-1} \mathbf{u}' \\ &= (1,0) \begin{bmatrix} 1-p & p \\ 1-p & 0 \end{bmatrix}^{t-1} \begin{bmatrix} 0 \\ p \end{bmatrix}, \end{aligned}$$

for $t = 2,3,....$ Then,

$$M(n) = 1-(1,0)\begin{bmatrix} 1-p & p \\ 1-p & 0 \end{bmatrix}^n \begin{bmatrix} 1 \\ 1 \end{bmatrix} + \sum_{t=1}^{n-1} M(n-t)(1,0)\begin{bmatrix} 1-p & p \\ 1-p & 0 \end{bmatrix}^{t-1} \begin{bmatrix} 0 \\ p \end{bmatrix},$$

for $n > 1$, and $M(1) = 0$. Let $p = 0.5$. Then,

$$\begin{aligned} M(1) &= 0 \\ M(2) &= 0.25 \\ M(3) &= 0.375 \\ M(4) &= 0.5625 \\ M(5) &= 0.7188 \\ M(10) &= 1.5557 \\ M(20) &= 3.2222. \end{aligned}$$

4.3 BIVARIATE PHASE-TYPE DISTRIBUTIONS

Let (T_1,T_2) have a discrete bivariate phase-type distribution with state space $\tilde{D} = \{0,1,...,d\}$, a transition probability matrix $\tilde{\mathbf{P}} = (\tilde{p}_{jk})_{0\le j,k\le d}$, and a starting distribution $\tilde{\pi} = (\pi_0,\pi_1,...,\pi_d) = (\pi_0,\pi)$. For an absorbing state $\{0\}$, define inaccessible subsets C_1 and C_2 such that $\tilde{D} = C_1 \cup C_2 \cup \{0\}$. For a discrete phase variable (T_1,T_2) having characteristics $(\tilde{D},\tilde{\mathbf{P}},\tilde{\pi},C)$ with $C = \{C_1,C_2\}$, the joint survival function is given by

$$P\{T_1 > t_1, T_2 > t_2\} = \pi\mathbf{P}^{t_1}\mathbf{\Delta}^{C_1}\mathbf{P}^{t_2-t_1}\mathbf{\Delta}^{C_2}\mathbf{e}', \tag{4.14}$$

$$P\{T_1 > t_2, T_2 > t_1\} = \pi\mathbf{P}^{t_1}\mathbf{\Delta}^{C_2}\mathbf{P}^{t_2-t_1}\mathbf{\Delta}^{C_1}\mathbf{e}', \tag{4.15}$$

for $t_1 < t_2$, where $\mathbf{P} = (\tilde{p}_{jk})_{1\le j,k\le d}$ is the sub-matrix of $\tilde{\mathbf{P}}$ suppressing the first line and the first column, $D = \{1,...,d\}$, $\mathbf{e} = (1,...,1)$ and $\mathbf{\Delta}^C \in \mathbb{R}^{d\times d}$ is defined by

$$(\mathbf{\Delta}^C)_{jk} = \begin{cases} 1, & \text{if } j = k \in C \\ 0 & \text{otherwise} \end{cases}$$

(Eisele 2008). The joint probability mass function of (T_1,T_2) is

$$\begin{aligned} P\{T_1 = T_2 = 0\} &= \pi_0, \\ P\{T_1 = 0, T_2 = t\} &= \boldsymbol{\pi}\mathbf{\Delta}^{D\setminus C_1}\mathbf{P}^{t-1}\mathbf{u}', \\ P\{T_1 = t, T_2 = 0\} &= \boldsymbol{\pi}\mathbf{\Delta}^{D\setminus C_2}\mathbf{P}^{t-1}\mathbf{u}', \\ P\{T_1 = t_1, T_2 = t_2\} &= \pi\mathbf{P}^{t_1-1}\mathbf{\Delta}^{C_1}\mathbf{P}\mathbf{\Delta}^{\mathbf{C}_2\setminus\mathbf{C}_1}\mathbf{P}^{t_2-t_1-1}\mathbf{u}',\ t_1 < t_2, \\ P\{T_1 = t_1, T_2 = t_2\} &= \pi\mathbf{P}^{t_2-1}\mathbf{\Delta}^{C_2}\mathbf{P}\mathbf{\Delta}^{\mathbf{C}_1\setminus\mathbf{C}_2}\mathbf{P}^{t_1-t_2-1}\mathbf{u}',\ t_2 < t_1, \\ P\{T_1 = T_2 = t\} &= \pi\mathbf{P}^{t-1}\mathbf{\Delta}^{\mathbf{C}_1\cap\mathbf{C}_2}\mathbf{u}', \end{aligned}$$

where $\mathbf{u}' = (\mathbf{I}-\mathbf{P})\mathbf{e}'$.

Example 4.17. Consider the bivariate geometric distribution given by (2.20). The joint distribution of (W_1, W_2) is

$$
\begin{aligned}
&P\{W_1 = t_1, W_2 = t_2\} \\
&= \begin{cases} p_{00}^{t_1-1} p_{10}(1-p_{01}-p_{11})^{t_2-t_1-1}(p_{01}+p_{11}), & \text{if} \quad t_1 < t_2 \\ p_{11}p_{00}^{t_1-1}, & \text{if} \quad t_1 = t_2 \\ p_{00}^{t_2-1} p_{01}(1-p_{10}-p_{11})^{t_1-t_2-1}(p_{10}+p_{11}), & \text{if} \quad t_1 > t_2. \end{cases}
\end{aligned}
$$

Then, (W_1, W_2) is phase-type with $\tilde{D} = \{1,2,3,4,5,6,7,8\}$, where the states $\{1,2,3\}$ are absorbing, $C_1 = \{5,6,7\}, C_2 = \{4,6,8\}$, and $\pi = (0,0,1,0,0)$, the transition matrix among the transient states $\{4,5,6,7,8\}$ is

$$
\mathbf{P} = \begin{bmatrix} 0 & 0 & 0 & 0 & 1-p_{01}-p_{11} \\ 0 & 0 & 0 & 1-p_{10}-p_{11} & 0 \\ p_{10} & p_{01} & p_{00} & 0 & 0 \\ 0 & 0 & 0 & 1-p_{10}-p_{11} & 0 \\ 0 & 0 & 0 & 0 & 1-p_{01}-p_{11} \end{bmatrix},
$$

and

$$
\mathbf{\Delta}^{C_1} = \begin{bmatrix} 0 & 0 & 0 & 0 & 0 \\ 0 & 1 & 0 & 0 & 0 \\ 0 & 0 & 1 & 0 & 0 \\ 0 & 0 & 0 & 1 & 0 \\ 0 & 0 & 0 & 0 & 0 \end{bmatrix}, \quad \mathbf{\Delta}^{C_2} = \begin{bmatrix} 1 & 0 & 0 & 0 & 0 \\ 0 & 0 & 0 & 0 & 0 \\ 0 & 0 & 1 & 0 & 0 \\ 0 & 0 & 0 & 0 & 0 \\ 0 & 0 & 0 & 0 & 1 \end{bmatrix}.
$$

For a discrete bivariate phase variable (T_1, T_2), (Eisele 2008) studied the joint distribution of

$$
S_1 = \sum_{i=1}^{T_1} Y_i^{(1)}, \quad S_2 = \sum_{i=1}^{T_2} Y_i^{(2)},
$$

when (T_1, T_2), $\left(Y_i^{(1)}\right)_{i\geq 1}$ and $\left(Y_i^{(2)}\right)_{i\geq 1}$ are mutually independent. (Tank and Eryilmaz 2020) obtained the joint distribution of (S_1, S_2) when there is a dependence between T_1 and $\left(Y_i^{(1)}\right)_{i\geq 1}$ and between T_2 and $\left(Y_i^{(2)}\right)_{i\geq 1}$. The model considered by (Tank and Eryilmaz 2020) is defined as follows. Let $(X_1^{(1)}, Y_1^{(1)}), (X_2^{(1)}, Y_2^{(1)}), \ldots$ be a sequence of independent and identically distributed random vectors with common probability mass function

$$
f_1(x,y) = P\left\{X_t^{(1)} = x, Y_t^{(1)} = y\right\}, \ t = 1,2,\ldots,
$$

and $(X_1^{(2)}, Y_1^{(2)}), (X_2^{(2)}, Y_2^{(2)}), \ldots$ be a sequence of independent and identically distributed random vectors with common probability mass function

$$
f_2(x,y) = P\left\{X_t^{(2)} = x, Y_t^{(2)} = y\right\}, \ t = 1,2,\ldots
$$

The random variable T_k is defined on a binary sequence $I_1^{(k)}, I_2^{(k)}, \ldots$ of random variables such that

$$I_t^{(k)} = \begin{cases} 1, & \text{if } X_t^{(k)} \in R_k \\ 0, & X_t^{(k)} \notin R_k, \end{cases}$$

for $k = 1,2$, where $R_1, R_2 \subset [0,\infty)$ are two critical regions. As note by (Tank and Eryilmaz 2020), the dependence between T_1 and T_2 can be created by the dependence between $X_t^{(1)}$ and $X_t^{(2)}$ for all t. Assume that (T_1, T_2) is a bivariate phase variable having characteristics $(\tilde{D}, \tilde{\mathbf{P}}, \tilde{\pi}, C)$ and following the joint survival function (4.14) and (4.15) with $P\{T_1 > 0, T_2 > 0\} = 1$. Define the following conditional probability generating functions:

$$\begin{aligned}
P_1(z_1,z_2) &= E(z_1^{Y^{(1)}} z_2^{Y^{(2)}} \mid X^{(1)} \in R_1, X^{(2)} \in R_2),\\
P_2(z_1,z_2) &= E(z_1^{Y^{(1)}} z_2^{Y^{(2)}} \mid X^{(1)} \in R_1, X^{(2)} \notin R_2),\\
P_3(z_1,z_2) &= E(z_1^{Y^{(1)}} z_2^{Y^{(2)}} \mid X^{(1)} \notin R_1, X^{(2)} \in R_2),\\
P_4(z_1,z_2) &= E(z_1^{Y^{(1)}} z_2^{Y^{(2)}} \mid X^{(1)} \notin R_1, X^{(2)} \notin R_2),
\end{aligned}$$

and

$$\begin{aligned}
U_1(z) &= E(z^{Y^{(1)}} \mid X^{(1)} \in R_1),\\
U_2(z) &= E(z^{Y^{(1)}} \mid X^{(1)} \notin R_1),\\
V_1(z) &= E(z^{Y^{(2)}} \mid X^{(2)} \in R_2),\\
V_2(z) &= E(z^{Y^{(2)}} \mid X^{(2)} \notin R_2).
\end{aligned}$$

Then, (Tank and Eryilmaz 2020) obtained the following expression for the joint probability generating function of S_1 and S_2:

$$\begin{aligned}
E(z_1^{S_1} z_2^{S_2}) &= \pi(\mathbf{I} - \mathbf{P}^{(1)}_{z_1,z_2})^{-1} \mathbf{\Delta}^{C_2} \mathbf{P}^{(1)}_{z_1,z_2} \mathbf{\Delta}^{C_1 \setminus C_2} (\mathbf{I} - \mathbf{P}^{(1)}_{z_1,z_2})^{-1} \mathbf{u}^{(1)}_{z_1,z_2}\\
&+ \pi(\mathbf{I} - \mathbf{P}^{(2)}_{z_1,z_2})^{-1} \mathbf{\Delta}^{C_1} \mathbf{P}^{(2)}_{z_1,z_2} \mathbf{\Delta}^{C_2 \setminus C_1} (\mathbf{I} - \mathbf{P}^{(2)}_{z_1,z_2})^{-1} \mathbf{u}^{(2)}_{z_1,z_2}\\
&+ \pi(\mathbf{I} - \mathbf{P}^{(3)}_{z_1,z_2})^{-1} \mathbf{\Delta}^{C_1 \cap C_2} \mathbf{u}^{(3)}_{z_1,z_2},
\end{aligned}$$

where $\mathbf{P}^{(1)}_{z_1,z_2}, \mathbf{u}^{(1)}_{z_1,z_2}, \mathbf{P}^{(2)}_{z_1,z_2}, \mathbf{u}^{(2)}_{z_1,z_2}, \mathbf{P}^{(3)}_{z_1,z_2}, \mathbf{u}^{(3)}_{z_1,z_2}$ are obtained by the following replacements of the entries of $\mathbf{P}$ and $\mathbf{u}' = (\mathbf{I} - \mathbf{P})\mathbf{e}'$. Note that $a \longrightarrow b$ represents the replacement of the term a by the term b.

$\mathbf{P};\mathbf{u}$		$\mathbf{P}^{(1)}_{z_1,z_2}, \mathbf{u}^{(1)}_{z_1,z_2}$	$\mathbf{P}^{(2)}_{z_1,z_2}, \mathbf{u}^{(2)}_{z_1,z_2}$	$\mathbf{P}^{(3)}_{z_1,z_2}, \mathbf{u}^{(3)}_{z_1,z_2}$
p_{11}	$\longrightarrow$	$p_{11}P_1(z_1,z_2)$	$p_{11}P_1(z_1,z_2)$	$p_{11}P_1(z_1,z_2)$
p_{12}	$\longrightarrow$	$p_{12}P_2(z_1,z_2)$	$p_{12}P_2(z_1,z_2)$	$p_{12}P_2(z_1,z_2)$
p_{21}	$\longrightarrow$	$p_{21}P_3(z_1,z_2)$	$p_{21}P_3(z_1,z_2)$	$p_{21}P_3(z_1,z_2)$
p_{22}	$\longrightarrow$	$p_{22}P_4(z_1,z_2)$	$p_{22}P_4(z_1,z_2)$	$p_{22}P_4(z_1,z_2)$
p_1	$\longrightarrow$	$p_1U_1(z_1)$	$p_1V_1(z_2)$	
$1-p_1$	$\longrightarrow$	$(1-p_1)U_2(z_1)$	$(1-p_2)V_2(z_2)$	

REFERENCES

1. Alfa, A.S. 2016. *Applied Discrete-Time Queues*. Springer, New York.
2. Alkaff, A. and Qomarudin, M.N. 2020. Modeling and analysis of system reliability using phase-type distribution closure properties. *Applied Stochastic Models in Business and Industry* 36: 548-569.
3. Balakrishnan, N., and Chan, P.S. 2000. Start-up demonstration tests with rejection of units upon observing d failures. *Annals of the Institute of Statistical Mathematics* 52: 184-196.
4. Eisele, K.T. 2006. Recursions for compound phase distributions. *Insurance: Mathematics Economics* 38: 149-156.
5. Eisele, K.T. 2008. Recursions for multivariate compound phase variables. *Insurance: Mathematics Economics* 42: 657-672.
6. Eryilmaz, S. 2019. Statistical inference for a class of startup demonstration tests. *Journal of Quality Technology* 51: 314-324.
7. Eryilmaz, S. and Tekin, M. 2019. Reliability evaluation of a system under a mixed shock model. *Journal of Computational and Applied Mathematics* 352: 255-261.
8. Eryilmaz, S. 2021. Revisiting discrete time age replacement policy for phase-type lifetime distributions. *European Journal of Operational Research*, 295: 699-704.
9. Kan, C. and Eryilmaz, S. 2021. Reliability assessment of a discrete time cold standby repairable system, *TOP* 29: 613-628.
10. Koutras, M.V. 1996. On a Markov chain approach for the study of reliability structures. *Journal of Applied Probability* 33: 357-367.
11. Koutras, M.V. and Eryilmaz, S. 2017. Compound geometric distribution of order k. *Methodology and Computing in Applied Probability* 19: 377-393.
12. Koutras, M.V. and Rakitzis, A.C. 2019. Control charts with random interarrival times between successive samplings. *Applied Stochastic Models in Business and Industry* 35: 651-670.
13. Lian, Z., Liu, L. and Neuts, M.F. 2005. A discrete-time model for common lifetime inventory systems. *Mathematics of Operations Research* 30: 718-732.
14. Mallor, F., and Omey, E. 2001. Shocks, runs and random sums. *Journal of Applied Probability* 38: 438-448.
15. Montoro-Cazorla, D. and Perez-Ocon, R. 2010. System availability in a shock model under preventive repair and phase-type distributions, *Applied Stochastic Models in Business and Industry* 26: 689-704.
16. Neuts, M.F. 1981. *Matrix-Geometric Solutions in Stochastic Models: An Algorithmic Approach*. The Johns Hopkins University Press, Baltimore.
17. Neuts, M. F. and Meier, K. S. 1981. On the use of phase-type distributions in reliability modeling of systems with two components. *OR Spektrum* 2: 227-234.
18. Tank, F. and Eryilmaz, S. 2015. The distributions of sum, minima and maxima of generalized geometric random variables. *Statistical Papers* 56: 1191-1203.
19. Tank, F. and Eryilmaz, S. 2020. On bivariate compound sums. *Journal of Computational and Applied Mathematics* 365: 112371.
20. Zhao, X., Cai, K., Wang, X. and Song, Y. 2018. Optimal replacement policies for a shock model with a change point. *Computers and Industrial Engineering* 118: 383-393.

5 Matrix-geometric Distributions

In this section, we consider a class of matrix-geometric distributions which is larger than the class of discrete phase-type distributions. Although the probability mass function of matrix-geometric distribution has the same form as discrete phase-type distributions, matrix-geometric distributions do not necessarily admit the Markov chain based probabilistic interpretation of discrete phase-type distributions. First, the definition and some properties of matrix-geometric distributions are summarized. Then, the usefulness of matrix-geometric distributions is demonstrated for the reliability evaluation of cold standby repairable system.

5.1 DEFINITION AND CHARACTERISTICS

A random variable T is said to have a matrix-geometric distribution if its probability mass function has the form

$$P\{T=t\} = \pi \mathbf{Q}^{t-1}\mathbf{u}', \tag{5.1}$$

for $t = 1,2,...,$ where $\mathbf{Q}$ is an m by m matrix, $\mathbf{u}'$ is a column vector of length m and π is a row vector of m. We shall use $T \sim MG(\pi, \mathbf{Q}, \mathbf{u})$ to represent that the random variable T has a matrix-geometric distribution with parameters π, $\mathbf{Q}$, and $\mathbf{u}$. Throughout the chapter we will assume that $P\{T=0\} = 0$.

If the random variable T has matrix-geometric distribution, then it has a rational probability generating function which is given by

$$\psi(z) = E(z^T) = \frac{c_1 z + ... + c_m z^m}{1 + d_1 z + ... + d_m z^m} \tag{5.2}$$

for some $m \geq 1$ and real constants $c_1,...,c_m$ and $d_1,...,d_m$. Using (5.2), the probability mass function and survival function of T can be represented as

$$P\{T=t\} = \pi \mathbf{Q}^{t-1}\mathbf{u}',$$

and

$$P\{T>t\} = \pi \mathbf{Q}^{t}(\mathbf{I}-\mathbf{Q})^{-1}\mathbf{u}', \tag{5.3}$$

where $\pi = (1,0,...,0)$,

$$\mathbf{Q} = \begin{bmatrix} -d_1 & 0 & 0 & \cdots & 0 & 1 \\ -d_m & 0 & 0 & \cdots & 0 & 0 \\ -d_{m-1} & 1 & 0 & \cdots & 0 & 0 \\ -d_{m-2} & 0 & 1 & & 0 & 0 \\ \vdots & \vdots & \vdots & \ddots & \vdots & \vdots \\ -d_2 & 0 & 0 & & 1 & 0 \end{bmatrix}, \quad \mathbf{u}' = \begin{bmatrix} c_1 \\ c_m \\ c_{m-1} \\ c_{m-2} \\ \vdots \\ c_2 \end{bmatrix}$$

DOI: 10.1201/9781003160632-5

(see, e.g. Bladt and Nielsen 2017). The expected value of T can be computed from

$$E(T) = \pi(\mathbf{I}-\mathbf{Q})^{-2}\mathbf{u}'.$$

Example 5.1. Consider the system that has been previously defined in Example 4.2. The system consists of n linearly ordered components labeled from 1 to n, and it fails if there exist at least f failed components or at least k consecutive failed components. The components can be in one of two states either working (1) or failed (0), and during a fixed period of time each component works with probability p. Assume that $f = 3$ and $k = 2$. Let T be the number of trials until getting either $f = 3$ failures or $k = 2$ consecutive failures in a sequence of Bernoulli trials with the probability of success p. Then, the reliability of the system can be expressed as $R = P\{T > n\}$. The probability generating function of T is

$$\psi(z) = \frac{(1-p)^2z^2 - 2p(1-p)^2z^3 + p(1-p)^2z^4}{1-3pz+3p^2z^2-p^3z^3} \tag{5.4}$$

(Triantafyllou and Koutras 2014). Clearly, (5.4) has the same form with (5.2). Therefore, T has matrix-geometric distribution with representation $T \sim MG(\pi,\mathbf{Q},\mathbf{u})$ with $\pi = (1,0,0,0)$,

$$\mathbf{Q} = \begin{bmatrix} 3p & 0 & 0 & 1 \\ 0 & 0 & 0 & 0 \\ p^3 & 1 & 0 & 0 \\ -3p^2 & 0 & 1 & 0 \end{bmatrix}, \quad \mathbf{u}' = \begin{bmatrix} 0 \\ p(1-p)^2 \\ -2p(1-p)^2 \\ (1-p)^2 \end{bmatrix}.$$

Thus, for given n, the reliability of the system can be computed using (5.3). ■

Example 5.2. For the *Geo*/*Geo*/1 system previously studied in Example 3.2, let T be the waiting time in the system for a machine (customer). Then, the probability generating function of T is

$$\psi(z) = \left(1-\frac{p}{d}\right)\frac{(dz+(1-p)(1-z))dz}{((1-p)(1-z)+z(d-p))(1-(1-d)z)}$$

(Alfa 2016). After some simple manipulations, we can obtain

$$\psi(z) = \frac{c_1z+c_2z^2}{1+d_1z+d_2z^2},$$

where

$$c_1 = \frac{d+p^2-p(1+d)}{1-p}, \quad c_2 = \frac{d^2-d-p^2+p}{1-p},$$

$$d_1 = \frac{p-2+d(2-p)}{1-p}, \quad d_2 = \frac{(1-d)^2}{1-p}.$$

Thus, the waiting time in the *Geo*/*Geo*/1 system can be represented as a matrix-geometric distribution with representation $T \sim MG(\pi,\mathbf{Q},\mathbf{u})$ with $\pi = (1,0)$,

$$\mathbf{Q} = \begin{bmatrix} -d_1 & 1 \\ -d_2 & 0 \end{bmatrix}, \quad \mathbf{u}' = \begin{bmatrix} c_1 \\ c_2 \end{bmatrix}. \quad ■$$

5.2 PROPERTIES

Matrix-geometric distributions have similar properties as phase-type distributions. Below, we present some of the properties of matrix-geometric distributions. The proofs can be found in (Neuts 1975). See also (Campillo Navarro 2019).

Let T_1 and T_2 be two independent random variables such that $T_1 \sim MG(\pi_1, \mathbf{Q}_1, \mathbf{u}_1)$ and $T_2 \sim MG(\pi_2, \mathbf{Q}_2, \mathbf{u}_2)$.

Property 1. $\min(T_1, T_2) \sim MG(\gamma, \mathbf{L}, \mathbf{v})$, where

$$\gamma = \pi_1 \otimes \pi_2,$$

and

$$\mathbf{L} = \mathbf{Q}_1 \otimes \mathbf{Q}_2 \ , \ \mathbf{v}' = (\mathbf{I} - (\mathbf{Q}_1 \otimes \mathbf{Q}_2))\,\mathbf{e}'$$

where $\otimes$ denotes Kronecker product.

Property 2. $\max(T_1, T_2) \sim MG(\gamma, \mathbf{L}, \mathbf{v})$, where

$$\gamma = (-(\pi_1 \otimes \pi_2), \pi_1, \pi_2),$$

and

$$\mathbf{L} = \begin{bmatrix} \mathbf{Q}_1 \otimes \mathbf{Q}_2 & \mathbf{0} & \mathbf{0} \\ \mathbf{0} & \mathbf{Q}_1 & \mathbf{0} \\ \mathbf{0} & \mathbf{0} & \mathbf{Q}_2 \end{bmatrix} \ , \ \mathbf{v}' = \mathbf{e}' - \mathbf{L}\mathbf{e}'.$$

Property 3. $T_1 + T_2 \sim MG(\gamma, \mathbf{L}, \mathbf{v})$, where

$$\gamma = (\pi_1, \mathbf{0}),$$

and

$$\mathbf{L} = \begin{bmatrix} \mathbf{Q}_1 & \mathbf{u}_1'\pi_2 \\ \mathbf{0} & \mathbf{Q}_2 \end{bmatrix} \ , \ \mathbf{v}' = \begin{bmatrix} \mathbf{0} \\ \mathbf{u}_2' \end{bmatrix}.$$

Property 4. Let N be an integer valued random variable having representation $N \sim MG(\pi, \mathbf{Q}_1, \mathbf{u})$ and $Y_1, Y_2, \ldots$ be a sequence of independent and identically distributed random variables such that $Y_i \sim MG(\sigma, \mathbf{Q}_2, \mathbf{v}), i = 1, 2, \ldots$, and N and $Y_1, Y_2, \ldots$ are independent, then

$$\sum_{i=1}^{N} Y_i \sim MG(\gamma, \mathbf{L}, \mathbf{w}), \tag{5.5}$$

where

$$\gamma = \sigma \otimes \pi, \tag{5.6}$$

and

$$\mathbf{L} = \mathbf{Q}_2 \otimes \mathbf{I} + (\mathbf{v}'\sigma) \otimes \mathbf{Q}_1 \ , \ \mathbf{w}' = \mathbf{v}' \otimes \mathbf{u}'. \tag{5.7}$$

Example 5.3. A series system consists of two components which are subjected to shocks over time. Assume that neither of the components are affected by the shock with probability p_0, the first component is affected but the second is not with probability p_1, the second component is affected but the first is not with probability p_2, and both components are affected by the shock with probability p_3, where $p_0 + p_1 + p_2 + p_3 = 1$. Suppose that component 1 and component 2 fail when they are affected by two consecutive shocks (Eryilmaz 2019). Let N_i denote the number of shocks until failure of the component i, $i = 1,2$. Then, the lifetime of the system corresponds to

$$T = \sum_{i=1}^{\min(N_1,N_2)} Y_i,$$

where $Y_1, Y_2, \ldots$ is a sequence of random variables that represent the times between successive shocks. The probability generating function of the random variable $\min(N_1,N_2)$ is

$$\psi(z) = \frac{c_1 z + c_2 z^2 + c_3 z^3 + c_4 z^4}{1 + d_1 z + d_2 z^2 + d_3 z^3 + d_4 z^4},$$

where

$$c_1 = 0\ ,\ c_2 = (p_1 + p_2 + p_3)(1 - p_0) - 2p_1 p_2,$$
$$c_3 = 2p_1 p_2 - 2p_0 p_1 p_2 - p_1 p_2^2 - p_1^2 p_2\ ,\ c_4 = p_1 p_2 p_3 (p_0 - 1),$$

and

$$d_1 = -p_0\ ,\ d_2 = -p_0(p_1 + p_2 + p_3) - p_1 p_2,$$
$$d_3 = -p_0 p_1 p_2\ ,\ d_4 = p_0 p_1 p_2 p_3.$$

Thus, $\min(N_1,N_2) \sim MG(\pi, \mathbf{Q}, \mathbf{u})$, where $\pi = (1,0,0,0)$,

$$\mathbf{Q} = \begin{bmatrix} -d_1 & 0 & 0 & 1 \\ -d_4 & 0 & 0 & 0 \\ -d_3 & 1 & 0 & 0 \\ -d_2 & 0 & 1 & 0 \end{bmatrix},\ \mathbf{u}' = \begin{bmatrix} 0 \\ c_4 \\ c_3 \\ c_2 \end{bmatrix}.$$

Let $P\{Y_i = y\} = \theta(1-\theta)^{y-1}, y = 1,2,\ldots$. Then, from (5.5)-(5.7), we obtain $T \sim MG(\gamma, \mathbf{L}, \mathbf{w})$, where $\gamma = (1,0,0,0)$,

$$\mathbf{L} = (1-\theta) \otimes \mathbf{I} + (\theta \otimes \mathbf{Q}),$$

and $\mathbf{w}' = \theta \mathbf{u}'$.

Example 5.4. For a sequence $Z_1, Z_2, \ldots$ of binary trials with two possible outcomes denoted by $\{0,1\}$, let E be the that event that at least k_1 consecutive 0s are followed by at least k_2 consecutive 1s. Denote by T_r the number of trials for the rth occurrence of the event E in $Z_1, Z_2, \ldots$. This waiting time random variable has been found to be useful in meteorology and agriculture. The probability generating function of T_r has been obtained by (Dafnis et al. 2010):

$$\psi(z) = \left(\frac{q^{k_1} p^{k_2} z^{k_1+k_2}}{1 - z + q^{k_1} p^{k_2} z^{k_1+k_2}} \right)^r, \tag{5.8}$$

where $p = P\{Z_t = 1\}$ and $q = P\{Z_t = 0\} = 1 - p, t \geq 1$. For $r = 1, T_1 \sim MG(\pi, \mathbf{Q}, \mathbf{u})$ with $\pi = (1, 0, ..., 0)$,

$$\mathbf{Q} = \begin{bmatrix} 1 & 0 & 0 & \cdots & 0 & 1 \\ -q^{k_1}p^{k_2} & 0 & 0 & \cdots & 0 & 0 \\ 0 & 1 & 0 & \cdots & 0 & 0 \\ 0 & 0 & 1 & \cdots & 0 & 0 \\ \vdots & & & \ddots & & \\ 0 & 0 & 0 & \cdots & 1 & 0 \end{bmatrix}, \quad \mathbf{u}' = \begin{bmatrix} 0 \\ q^{k_1}p^{k_2} \\ 0 \\ 0 \\ \vdots \\ 0 \end{bmatrix}. \tag{5.9}$$

The random variable T_r that has the probability generating function (5.8) can be represented as the sum of r independent random variables each have matrix-geometric distribution. That is,

$$T_r = \sum_{i=1}^{r} T_1^{(i)},$$

where $T_1^{(1)}, T_1^{(2)}, ..., T_1^{(r)}$ are independent and $T_1^{(i)} \sim MG(\pi, \mathbf{Q}, \mathbf{u})$, where $\mathbf{Q}$ and $\mathbf{u}$ are given by (5.9). Thus, for given r, the distribution of T_r can be obtained via Property 3. Indeed, for $T_1^{(1)} \sim MG(\pi, \mathbf{Q}, \mathbf{u})$ and $T_1^{(2)} \sim MG(\pi, \mathbf{Q}, \mathbf{u})$, we have $T_1^{(1)} + T_1^{(2)} \sim MG(\pi_2, \mathbf{Q}_2, \mathbf{u}_2)$, where $\pi_2 = (\pi, \mathbf{0}_k)$, and

$$\mathbf{Q}_2 = \begin{bmatrix} \mathbf{Q} & \mathbf{u}'\pi \\ \mathbf{0}_{k\times k} & \mathbf{Q} \end{bmatrix}_{2k\times 2k}, \quad \mathbf{u}_2' = \begin{bmatrix} \mathbf{0}_k \\ \mathbf{u}' \end{bmatrix}_{2k\times 1},$$

where $k = k_1 + k_2$, $\mathbf{0}_k$ is a vector of zeros, and $\mathbf{0}_{k\times k}$ is a matrix of zeros. Thus, for $r > 1$, we can obtain

$$T_r \sim MG(\pi_r, \mathbf{Q}_r, \mathbf{u}_r),$$

where $\pi_r = (\pi_{r-1}, \mathbf{0})$, and

$$\mathbf{Q}_r = \begin{bmatrix} \mathbf{Q}_{r-1} & \mathbf{u}_{r-1}'\pi \\ \mathbf{0} & \mathbf{Q} \end{bmatrix}_{rk\times rk}, \quad \mathbf{u}_r' = \begin{bmatrix} \mathbf{0} \\ \mathbf{u}' \end{bmatrix}_{rk\times 1},$$

with $\pi_1 = \pi, \mathbf{Q}_1 = \mathbf{Q}$, and $\mathbf{u}_1 = \mathbf{u}$. For an illustration, let us find the distribution of T_2 when $k_1 = 2$ and $k_2 = 3$. Clearly, $MG(\pi, \mathbf{Q}, \mathbf{u})$ with $\pi = (1, 0, 0, 0, 0)$,

$$\mathbf{Q} = \begin{bmatrix} 1 & 0 & 0 & 0 & 1 \\ -q^2p^3 & 0 & 0 & 0 & 0 \\ 0 & 1 & 0 & 0 & 0 \\ 0 & 0 & 1 & 0 & 0 \\ 0 & 0 & 0 & 1 & 0 \end{bmatrix}, \quad \mathbf{u}' = \begin{bmatrix} 0 \\ q^2p^3 \\ 0 \\ 0 \\ 0 \end{bmatrix}.$$

Then, $T_2 \sim MG(\pi_2, \mathbf{Q}_2, \mathbf{u}_2)$ with $\pi_2 = (1,0,0,0,0,0,0,0,0,0)$,

$$\mathbf{Q}_2 = \begin{bmatrix} 1 & 0 & 0 & 0 & 1 & 0 & 0 & 0 & 0 & 0 \\ -q^2p^3 & 0 & 0 & 0 & 0 & q^2p^3 & 0 & 0 & 0 & 0 \\ 0 & 1 & 0 & 0 & 0 & 0 & 0 & 0 & 0 & 0 \\ 0 & 0 & 1 & 0 & 0 & 0 & 0 & 0 & 0 & 0 \\ 0 & 0 & 0 & 1 & 0 & 0 & 0 & 0 & 0 & 0 \\ 0 & 0 & 0 & 0 & 0 & 1 & 0 & 0 & 0 & 1 \\ 0 & 0 & 0 & 0 & 0 & -q^2p^3 & 0 & 0 & 0 & 0 \\ 0 & 0 & 0 & 0 & 0 & 0 & 1 & 0 & 0 & 0 \\ 0 & 0 & 0 & 0 & 0 & 0 & 0 & 1 & 0 & 0 \\ 0 & 0 & 0 & 0 & 0 & 0 & 0 & 0 & 1 & 0 \end{bmatrix},$$

and $\mathbf{u}_2 = (0,0,0,0,0,0,q^2p^3,0,0,0)$. The mean of T_2 is found to be

$$E(T_2) = \pi_2(\mathbf{I} - \mathbf{Q}_2)^{-2}\mathbf{u}_2' = \frac{2}{p^3q^2}.$$

Example 5.5. Consider the linear consecutive-2-out-of-n:F system that consists of n linearly ordered components, and fails if there exists at least two consecutive failed components. Assume that the first n_1 components have common reliability p_1 and the remaining $n - n_1$ components have common reliability p_2. If the random variable ξ denotes the number of trials until the occurrence of two consecutive failures (zeros) in a sequence of binary trials, then the reliability of the linear consecutive-2-out-of-n:F system can be written as

$$P\{\xi > n\}.$$

For $n_1 = 2$, the probability generating function of ξ has been obtained by (Eryilmaz and Tank 2021):

$$\psi(z) = \frac{(1-p_1)^2z^2 + (1-p_1)(p_1-p_2)z^3 + (1-p_2)(p_1-p_2)z^4}{1 - p_2z + p_2(p_2-1)z^2}.$$

Thus, the reliability of the linear consecutive-2-out-of-n:F system can be computed from

$$P\{\xi > n\} = \pi\mathbf{Q}^n(\mathbf{I} - \mathbf{Q})^{-1}\mathbf{u}', \tag{5.10}$$

where $\pi = (1,0,0,0)$,

$$\mathbf{Q} = \begin{bmatrix} p_2 & 0 & 0 & 1 \\ 0 & 0 & 0 & 0 \\ 0 & 1 & 0 & 0 \\ p_2(1-p_2) & 0 & 1 & 0 \end{bmatrix},$$

and

$$\mathbf{u}' = \begin{bmatrix} 0 \\ (1-p_2)(p_1-p_2) \\ (1-p_1)(p_1-p_2) \\ (1-p_1)^2 \end{bmatrix}.$$

For an illustration, let $p_1 = 0.97$, $p_2 = 0.95$, and $n = 20$. Then, from (5.10), the reliability of the linear consecutive-2-out-of-n:F system is found to be 0.9579.

Example 5.6. Let W_r denote the number of trials until r successes (ones) in a homogeneous $\{0,1\}$-valued Markov chain $X_1, X_2, \ldots$ whose transition probability matrix is

$$\mathbf{P} = \begin{bmatrix} p_{00} & p_{01} \\ p_{10} & p_{11} \end{bmatrix},$$

and initial probabilities are $p_0 = P\{X_1 = 0\}$, $p_1 = P\{X_1 = 1\} = 1 - p_0$. (Viveros et al. 1984) obtained the following equation for the probability generating function of W_r:

$$\psi(z) = \left[p_1 z + \frac{(1-p_1)p_{01}z^2}{1-(1-p_{01})z}\right]\left[p_{11}z + \frac{(1-p_{11})p_{01}z^2}{1-(1-p_{01})z}\right]^{r-1}. \tag{5.11}$$

Let $r = 2$. Then (5.11) becomes

$$\begin{aligned}\psi(z) &= \left[p_1 z + \frac{(1-p_1)p_{01}z^2}{1-(1-p_{01})z}\right]\left[p_{11}z + \frac{(1-p_{11})p_{01}z^2}{1-(1-p_{01})z}\right] \\ &= \frac{1}{1+(2p_{01}-2)z+(1-p_{01})^2z^2}\left[p_1p_{11}z^2\right. \\ &\quad \left. + (p_1p_{01} - 2p_1p_{11} + p_{01}p_{11})z^3 + (p_{01}^2 - p_1p_{01} + p_1p_{11} - p_{01}p_{11})z^4\right].\end{aligned}$$

Therefore, W_2 has matrix-geometric distribution with representation $W_2 \sim MG(\pi, \mathbf{Q}, \mathbf{u})$ with $\pi = (1,0,0,0)$,

$$\mathbf{Q} = \begin{bmatrix} 2(1-p_{01}) & 0 & 0 & 1 \\ 0 & 0 & 0 & 0 \\ 0 & 1 & 0 & 0 \\ -(1-p_{01})^2 & 0 & 1 & 0 \end{bmatrix},$$

and

$$\mathbf{u}' = \begin{bmatrix} 0 \\ p_{01}^2 - p_1p_{01} + p_1p_{11} - p_{01}p_{11} \\ p_1p_{01} - 2p_1p_{11} + p_{01}p_{11} \\ p_1p_{11} \end{bmatrix}.$$

(Eryilmaz 2016) obtained the following alternative expression for the probability mass function of W_2:

$$P\{W_2 = t\} = (1,0,0)\begin{bmatrix} p_0 & 0 & p_1 \\ 0 & 0 & 0 \\ 0 & 0 & 0 \end{bmatrix}\begin{bmatrix} p_{00} & 0 & p_{01} \\ 0 & p_{00} & 0 \\ 0 & p_{10} & 0 \end{bmatrix}^{t-2}\begin{bmatrix} 0 \\ p_{01} \\ p_{11} \end{bmatrix},$$

for $t = 2, 3, \ldots$. The mean number of trials until two successes can be computed from

$$\begin{aligned}E(W_2) &= \pi(\mathbf{I}-\mathbf{Q})^{-2}\mathbf{u}' \\ &= \frac{2(1+p_{01}) - p_1 - p_{11}}{p_{01}}.\end{aligned}$$

For $p_1 = 0.7, p_{11} = 0.9$, and $p_{01} = 0.3$, we compute $P\{W_2 = t\}$ for selected values of t:

$$\begin{aligned} P\{W_2 = 2\} &= 0.6300, \\ P\{W_2 = 3\} &= 0.1020, \\ P\{W_2 = 5\} &= 0.0538, \\ P\{W_2 = 10\} &= 0.0106. \end{aligned}$$

It should be noted in the special case when $p_1 = p_{11} = p_{01} = p$, the random variable W_2 corresponds to the number of trials until two successes in a sequence of classical Bernoulli trials. That is, W_2 has negative Binomial distribution with mean

$$E(W_2) = \frac{2}{p}.$$

Example 5.7. Consider the negative Binomial distribution of order k which has been previously defined in section 2. The probability generating function of the corresponding random variable is

$$E(z^{T_{r.k}}) = \left[\frac{(pz)^k(1-pz)}{1-z+(1-p)p^k z^{k+1}}\right]^r. \tag{5.12}$$

Let $k = 3$ and $r = 2$. Then, (5.12) becomes

$$E(z^{T_{2.3}}) = \frac{p^6z^6 - 2p^7z^7 + p^8z^8}{1 - 2z + z^2 + (2p^3 - 2p^4)z^4 + (2p^4 - 2p^3)z^5 + (p^8 - 2p^7 + p^6)z^8}.$$

Thus, $T_{2.3} \sim MG(\pi, \mathbf{Q}, \mathbf{u})$ with $\pi = (1,0,0,0,0,0,0,0)$,

$$\mathbf{Q} = \begin{bmatrix} 2 & 0 & 0 & 0 & 0 & 0 & 0 & 1 \\ -(p^8 - 2p^7 + p^6) & 0 & 0 & 0 & 0 & 0 & 0 & 0 \\ 0 & 1 & 0 & 0 & 0 & 0 & 0 & 0 \\ 0 & 0 & 1 & 0 & 0 & 0 & 0 & 0 \\ -(2p^4 - 2p^3) & 0 & 0 & 1 & 0 & 0 & 0 & 0 \\ -(2p^3 - 2p^4) & 0 & 0 & 0 & 1 & 0 & 0 & 0 \\ 0 & 0 & 0 & 0 & 0 & 1 & 0 & 0 \\ -1 & 0 & 0 & 0 & 0 & 0 & 1 & 0 \end{bmatrix},$$

and

$$\mathbf{u}' = \begin{bmatrix} 0 \\ p^8 \\ -2p^7 \\ p^6 \\ 0 \\ 0 \\ 0 \\ 0 \end{bmatrix}.$$

For an illustration, let $p = 0.7$. Then,

$$\begin{aligned}
P\{T_{2,3} = t\} &= 0 \text{ for } t < 6, \\
P\{T_{2,3} = 6\} &= 0.1176, \\
P\{T_{2,3} = 7\} &= 0.0706, \\
P\{T_{2,3} = 10\} &= 0.0781, \\
P\{T_{2,3} = 15\} &= 0.0464, \\
P\{T_{2,3} = 20\} &= 0.0214.
\end{aligned}$$

The expected value of $T_{2,3}$ is

$$\begin{aligned}
E(T_{2,3}) &= \pi(\mathbf{I}-\mathbf{Q})^{-2}\mathbf{u}' \\
&= 2\frac{1-p^3}{(1-p)p^3} \\
&= 12.7697. \quad \blacksquare
\end{aligned}$$

Consider a discrete renewal process when the times between events follow matrix-geometric distribution. From the equation (2.24), the renewal function is

$$M(n) = F(n) + \sum_{t=1}^{n-1} M(n-t)p_t.$$

Using (5.1) and (5.3), the renewal function becomes

$$M(n) = 1 - \pi\mathbf{Q}^n(\mathbf{I}-\mathbf{Q})^{-1}\mathbf{u}' + \sum_{t=1}^{n-1} M(n-t)\pi\mathbf{Q}^{t-1}\mathbf{u}'.$$

5.3 TWO-UNIT COLD STANDBY REPAIRABLE SYSTEM VIA MATRIX-GEOMETRIC DISTRIBUTIONS

In this section, we model and evaluate a two-unit standby repairable system whose continuous time version has been introduced and studied by (Kus and Eryilmaz 2021). We first describe the model. The system consists of two identical units, one active and one standby which are new at time $t = 0$. The standby unit is switched into operation when the active unit fails. The repair operation is immediately performed on a failed unit if the damage on the failed unit is less than or equal to a predetermined threshold γ. In other words, if the damage is greater than the threshold γ, then the catastophic failure occurs and the unit cannot be repaired. The notation that is necessary for the description of the model is listed below.

Y_{i+1} : The magnitude of the damage upon the failure of the unit, which lives $X_i, i = 0,1,2,...$

Z_{i+1} : The repair time of the unit, which lives $X_i, i = 0,1,2,...$

γ: The critical predetermined threshold for a catastrophic failure.

The system fails if either a damage size upon the failure of the active component is larger than γ or the repair time of the failed unit exceeds the lifetime of the active unit, whichever happens first. Thus, the lifetime of the system is

$$T = \sum_{i=0}^{N} X_i, \tag{5.13}$$

where the random variable N denotes the waiting time for the failure of the system and is defined as

$$\{N = n\} \text{ iff } \{X_1 \geq Z_1, Y_1 \leq \gamma, \ldots, X_{n-1} \geq Z_{n-1}, Y_{n-1} \leq \gamma,$$
$$(Y_n > \gamma \vee X_n < Z_n, Y_n \leq \gamma)\}.$$

(Kus and Eryilmaz 2021) has investigated the distribution of the lifetime random variable defined by (5.13) under the following assumptions:

1. The unit becomes as good as new after repair.
2. Switching time is negligible and ideal.
3. The random variables $X_0, X_1, \ldots$ are independent and identically distributed.
4. The random variables $Y_1, Y_2, \ldots$ are independent and identically distributed.
5. The random variables $Z_1, Z_2, \ldots$ are independent and identically distributed.
6. The random variables Y_i and Z_i are dependent for all i.
7. The random sequence $(X_0, X_1, \ldots)$ is independent of $(Y_1, Y_2, \ldots)$ and $(Z_1, Z_2, \ldots)$.

Assuming that all random variables involved in the model are discrete-valued, the probability generating function of the system's lifetime is computed as follows.

$$\begin{aligned}
E(z^T) &= \sum_{n=1}^{\infty} E\left(z^{X_0+X_1+\ldots+X_n} \mid N = n\right) P\{N = n\} \\
&= E(z^X) \sum_{n=1}^{\infty} E\left(z^{X_1+\ldots+X_n} \mid X_1 \geq Z_1, Y_1 \leq \gamma, \ldots\right. \\
&\left., X_{n-1} \geq Z_{n-1}, Y_{n-1} \leq \gamma, Y_n > \gamma\right) [P\{X \geq Z, Y \leq \gamma\}]^{n-1} P\{Y > \gamma\} \\
&+ E(z^X) \sum_{n=1}^{\infty} E\left(z^{X_1+\ldots+X_n} \mid X_1 \geq Z_1, Y_1 \leq \gamma, \ldots\right. \\
&\left., X_{n-1} \geq Z_{n-1}, Y_{n-1} \leq \gamma, X_n < Z_n, Y_n \leq \gamma\right) \\
&\times [P\{X \geq Z, Y \leq \gamma\}]^{n-1} P\{X < Z, Y \leq \gamma\}.
\end{aligned}$$

Under the model assumptions, the probability generating function becomes

$$\begin{aligned}
E(z^T) = E(z^X) &\sum_{n=1}^{\infty} \left[E(z^X \mid X \geq Z, Y \leq \gamma)\right]^{n-1} E(z^X \mid Y > \gamma) \\
&\times [P\{X \geq Z, Y \leq \gamma\}]^{n-1} P\{Y > \gamma\} \\
&+ E(z^X) \sum_{n=1}^{\infty} \left[E(z^X \mid X \geq Z, Y \leq \gamma)\right]^{n-1} \\
\times E(z^X \mid X < Z, Y \leq \gamma) &[P\{X \geq Z, Y \leq \gamma\}]^{n-1} P\{X < Z, Y \leq \gamma\}.
\end{aligned}$$

After some simple manipulations,

$$\begin{aligned} E(z^T) &= \frac{E(z^X)E(z^X \mid Y > \gamma)P\{Y > \gamma\}}{1 - E(z^X \mid X \geq Z, Y \leq \gamma)P\{X \geq Z, Y \leq \gamma\}} \\ &+ \frac{E(z^X)E(z^X \mid X < Z, Y \leq \gamma)P\{X < Z, Y \leq \gamma\}}{1 - E(z^X \mid X \geq Z, Y \leq \gamma)P\{X \geq Z, Y \leq \gamma\}} \\ &= \frac{E(z^X)}{1 - E(z^X \mid X \geq Z, Y \leq \gamma)P\{X \geq Z, Y \leq \gamma\}} \\ &\times \left[E(z^X \mid Y > \gamma)P\{Y > \gamma\}\right. \\ &\left.+E(z^X \mid X < Z, Y \leq \gamma)P\{X < Z, Y \leq \gamma\}\right]. \end{aligned} \tag{5.14}$$

Manifestly,

$$\begin{aligned} &E(z^X \mid X \geq Z, Y \leq \gamma) \\ &\quad = \frac{1}{P\{X \geq Z, Y \leq \gamma\}} \sum_x z^x H(\gamma, x) P\{X = x\}, \end{aligned} \tag{5.15}$$

and

$$\begin{aligned} E(z^X \mid X < Z, Y \leq \gamma) = \frac{1}{P\{X < Z, Y \leq \gamma\}} &\left[P\{Y \leq \gamma\} E(z^X)\right. \\ &\left.-\sum_x z^x H(\gamma, x) P\{X = x\}\right], \end{aligned} \tag{5.16}$$

where

$$H(y,z) = P\{Y_i \leq y, Z_i \leq z\}.$$

By substituting (5.15) and (5.16) in (5.14), the probability generating function of the system's lifetime is obtained as

$$E(z^T) = \frac{E(z^X)\left[E(z^X) - \sum_x z^x H(\gamma, x) P\{X = x\}\right]}{1 - \sum_x z^x H(\gamma, x) P\{X = x\}}.$$

In the particular case when Y_i and Z_i are independent for all i, then

$$E(z^T) = \frac{E(z^X)\left[E(z^X) - (1-\theta)\sum_x z^x G(x) P\{X = x\}\right]}{1 - (1-\theta)\sum_x z^x G(x) P\{X = x\}},$$

where $G(z) = P\{Z \leq z\}$ and $\theta = P\{Y > \gamma\}$.

In the following example, we obtain an explicit expression for the probability generating function and the distribution of the system's lifetime when the units have geometric lifetime and repair distributions.

Example 5.8. For the system that has the lifetime (5.13), let

$$P\{X_i = x\} = p(1-p)^{x-1},\ x = 1,2,...,$$

and

$$P\{Z = z\} = \mu(1-\mu)^{z-1},\ z = 1,2,.....$$

Then, $G(z) = 1-(1-\mu)^z$, and

$$E(z^X) = \frac{zp}{1-z(1-p)}.$$

On the other hand,

$$\begin{aligned}
&\sum_{x=1}^{\infty} z^x G(x) P\{X = x\} \\
&= \sum_{x=1}^{\infty} z^x \left[1-(1-\mu)^x\right] p(1-p)^{x-1} \\
&= \frac{zp}{1-z(1-p)} - \frac{zp(1-\mu)}{1-z(1-\mu)(1-p)}.
\end{aligned}$$

Thus, the probability generating function of the system is found to be

$$E(z^T) = \frac{A(z)}{B(z)}, \tag{5.17}$$

where

$$A(z) = p^2\left[\theta+(1-\theta)(1-\mu)\right]z^2 - \left[p^2(1-p)(1-\mu)\right]z^3,$$

and

$$\begin{aligned}
B(z) = {} & 1+\left[(1-p)(\mu-3)-p(1-\theta)\mu\right]z \\
& + \left[(1-p)^2(3-2\mu)+p(1-p)(1-\theta)\mu\right]z^2 \\
& - (1-p)^3(1-\mu)z^3.
\end{aligned}$$

The probability generating function (5.17) has the same form with (5.2). Thus, the system's lifetime has matrix-geometric distribution with representation $T \sim MG(\pi, \mathbf{Q}, \mathbf{u})$, where $\pi = (1,0,0)$,

$$\mathbf{Q} = \begin{bmatrix} -(1-p)(\mu-3)+p(1-\theta)\mu & 0 & 1 \\ (1-p)^3(1-\mu) & 0 & 0 \\ -(1-p)^2(3-2\mu)-p(1-p)(1-\theta)\mu & 1 & 0 \end{bmatrix},$$

and

$$\mathbf{u}' = \begin{bmatrix} 0 \\ -p^2(1-p)(1-\mu) \\ p^2\left[\theta+(1-\theta)(1-\mu)\right] \end{bmatrix}.$$

Thus, the survival function of the system can be computed from

$$P\{T > t\} = \pi \mathbf{Q}^t (\mathbf{I} - \mathbf{Q})^{-1} \mathbf{u}'.$$

The mean time to failure of the system is found to be

$$\begin{aligned} E(T) &= \pi (\mathbf{I} - \mathbf{Q})^{-2} \mathbf{u}' \\ &= \frac{\mu(1 - 2p + \theta) + 2p}{p^2(1-\mu) + p\mu\theta}. \end{aligned}$$

For an illustration, let $p = 0.1, \mu = 0.5$, and $\theta = 0.1$. Then, $E(T) = 65$, and

$$\begin{aligned} P\{T > 2\} &= 0.9945, \\ P\{T > 3\} &= 0.9864, \\ P\{T > 5\} &= 0.9665, \\ P\{T > 10\} &= 0.9087, \\ P\{T > 20\} &= 0.7846. \end{aligned}$$

Using (4.6), the distribution of the system's lifetime can be alternatively computed by the following recursive scheme:

$$\begin{aligned} P\{T = 1\} &= 0, \\ P\{T = 2\} &= c_2, \\ P\{T = 3\} &= c_3 - d_1 c_2, \end{aligned}$$

and for $n > 3$,

$$P\{T = n\} = -d_1 P\{T = n-1\} - d_2 P\{T = n-2\} - d_3 P\{T = n-3\},$$

where

$$\begin{aligned} c_2 &= p^2 \left[\theta + (1-\theta)(1-\mu)\right], \\ c_3 &= -p^2(1-p)(1-\mu), \end{aligned}$$

and

$$\begin{aligned} d_1 &= (1-p)(\mu - 3) - p(1-\theta)\mu, \\ d_2 &= \left[(1-p)^2(3 - 2\mu) + p(1-p)(1-\theta)\mu\right], \\ d_3 &= -(1-p)^3(1-\mu). \end{aligned}$$

Example 5.9. Using the methodology described in Section 4, under the assumptions of Example 5.8, the distribution of the system's lifetime can also be represented as

$$P\{T = t\} = \sum_{k=0}^{2} \omega_k \lambda_k^{t-1} (1 - \lambda_k), \tag{5.18}$$

where λ_0, λ_1, and λ_2 are real distinct roots of the characteristic polynomial

$$\lambda^3 + d_1\lambda^2 + d_2\lambda + d_3 = 0,$$

and

$$\omega_k = \frac{1}{(1-\lambda_k)\prod_{\substack{i=0\\ i\neq k}}^{2}(\lambda_k - \lambda_i)} \sum_{j=0}^{2}(-1)^j p(3-j) \sum_{\substack{0\le i_1<\ldots<i_j\le 2\\ i_j\neq k}}\cdots\sum \lambda_{i_1}\ldots\lambda_{i_j},$$

for $k = 0,1,2$, where $p(t) = P\{T = t\}$. For an illustration, let again $p = 0.1, \mu = 0.5$, and $\theta = 0.1$. Then,

$$\lambda_0 = 0.412,\ \lambda_1 = 0.9,\ \lambda_2 = 0.983,$$

and

$$\omega_0 = -0.0136,\ \omega_1 = -0.1111,\ \omega_2 = 1.1247.$$

Thus, the distribution of the system's lifetime can also be represented as

$$\begin{aligned} &P\{T = t\}\\ &= \omega_0\lambda_0^{t-1}(1-\lambda_0) + \omega_1\lambda_1^{t-1}(1-\lambda_1) + \omega_2\lambda_2^{t-1}(1-\lambda_2). \end{aligned}$$

Manifestly,

$$E(T) = \omega_0\frac{1}{1-\lambda_0} + \omega_1\frac{1}{1-\lambda_1} + \omega_2\frac{1}{1-\lambda_2}. \quad \blacksquare$$

In the following Example, we calculate the coefficients and the parameter values in (5.18) when $p = 0.1, \mu = 0.5$ for selected values of the parameter $\theta = P\{Y > \gamma\}$.

Example 5.10. Let $p = 0.1, \mu = 0.5$ in Example 5.8. Then,

$$\theta = 0.05 \Longrightarrow \begin{array}{c} (\lambda_0,\lambda_1,\lambda_2) = (0.4102, 0.9, 0.9873)\\ (\omega_0,\omega_1,\omega_2) = (-0.0141, -0.0526, 1.0667) \end{array}$$

$$\theta = 0.25 \Longrightarrow \begin{array}{c} (\lambda_0,\lambda_1,\lambda_2) = (0.4175, 0.9, 0.97)\\ (\omega_0,\omega_1,\omega_2) = (-0.0122, -0.3333, 1.3455) \end{array}$$

$$\theta = 0.5 \Longrightarrow \begin{array}{c} (\lambda_0,\lambda_1,\lambda_2) = (0.4274, 0.9, 0.9476)\\ (\omega_0,\omega_1,\omega_2) = (-0.0092, -1, 2.0092). \end{array} \quad \blacksquare$$

(Eryilmaz 2021) investigated the limiting behavior of the hazard rate of a system whose lifetime distribution has the form

$$P\{T = t\} = \sum_{k=0}^{m-1}\omega_k\lambda_k^{t-1}(1-\lambda_k). \tag{5.19}$$

The hazard rate corresponding to (5.19) is

$$r(t) = \frac{P\{T = t\}}{P\{T \ge t\}} = \frac{\sum_{k=0}^{m-1}\omega_k\lambda_k^{t-1}(1-\lambda_k)}{\sum_{k=0}^{m-1}\omega_k\lambda_k^{t-1}}.$$

Then, as proved by (Eryilmaz 2021),

$$\lim_{t \longrightarrow \infty} r(t) = \min(1-\lambda_0, 1-\lambda_1, ..., 1-\lambda_{m-1}).$$

Example 5.11. Let $p = 0.1, \mu = 0.5$ in Example 5.8. Then,

$$\theta = 0.05 \Longrightarrow \lim_{t \longrightarrow \infty} r(t) = 0.0127,$$
$$\theta = 0.25 \Longrightarrow \lim_{t \longrightarrow \infty} r(t) = 0.03,$$
$$\theta = 0.5 \Longrightarrow \lim_{t \longrightarrow \infty} r(t) = 0.0524.$$

REFERENCES

1. Alfa, A.S. 2016. *Applied Discrete-Time Queues*. Springer, New York.
2. Bladt, M. and Nielsen, B.F. 2017. *Matrix-exponential Distributions in Applied Probability*. New York: Springer.
3. Campillo Navarro, A. 2019. Order statistics and multivariate discrete phase-type distributions. DTU Compute. DTU Compute PHD-2018, vol. 492.
4. Dafnis, S.P., Antzoulakos, D.L. and Philippou, A.N. 2010. Distributions related to (k_1, k_2) events. *Journal of Statistical Planning and Inference* 140: 1691-1700.
5. Eryilmaz, S. 2016. Compound Markov negative binomial distribution. *Journal of Computational and Applied Mathematics* 292: 1-6.
6. Eryilmaz, S. 2019. Modeling systems with two dependent components under bivariate shock models. *Communications in Statistics-Simulation and Computation* 48: 1714-1728.
7. Eryilmaz, S. 2021. Revisiting discrete time age replacement policy for phase-type lifetime distributions. *European Journal of Operational Research*, 295: 699-704.
8. Eryilmaz, S. and Tank, F. 2021. Computing minimal signature of coherent systems through matrix-geometric distributions. *Journal of Applied Probability* 58: 621-636.
9. Kus, C. and Eryilmaz, S. 2021. Analysis of the two-unit cold standby repairable system with damage and repair time dependency via matrix-exponential distributions. *Quality Technology & Quantitative Management* 18: 771-786.
10. Triantafyllou, I.S. and Koutras, M.V. 2014. Reliability properties of (n, f, k) systems. *IEEE Transactions on Reliability* 63: 357-366.
11. Viveros, R., Balasubramanian, K. and Balakrishnan, N. 1994. Binomial and negative Binomial analogues under correlated Bernoulli trials. *The American Statistician* 48: 243-247.

6 Some Particular Models/Classes of Discrete Distributions

In this section, we consider some particular models/classes of discrete distributions. We first mention about order statistics of a sequence of discrete valued random variables. Then, the concept of system signature which in fact defines a discrete probability distribution is presented. Finally, power series class of distributions and Panjer's family of distributions which contain well-known discrete distributions are described.

6.1 ORDER STATISTICS

Order statistics play an important role in probability, stochastic modeling, and statistical inference. They are also useful in various applications including reliability engineering and statistical quality control. Indeed, for a system consisting of n components, if the lifetimes of the components are denoted by $T_1, ..., T_n$, then

$$T_{1:n} = \min(T_1, ..., T_n),$$

and

$$T_{n:n} = \max(T_1, ..., T_n)$$

represent respectively the lifetimes of series and parallel systems. Let $T_{r:n}$ denote the rth smallest among $T_1, ..., T_n$ when $T_1, ..., T_n$ are arranged in a non-decreasing order, $r = 1, ..., n$.

Let $T_1, ..., T_n$ be independent and identically distributed random variables with common cumulative distribution function $F(t) = P\{T_i \leq t\}, i = 1, ..., n$. Then, the cumulative distribution function of the rth order statistic is

$$P\{T_{r:n} \leq t\} = \sum_{i=r}^{n} \binom{n}{i} F^i(t)(1 - F(t))^{n-i},$$

for $r = 1, ..., n$ (David and Nagaraja 2003).

If the random variables $T_1, ..., T_n$ have common discrete distribution, then

$$\begin{aligned} P\{T_{r:n} = t\} &= P\{T_{r:n} \leq t\} - P\{T_{r:n} \leq t-1\} \\ &= \sum_{i=r}^{n} \binom{n}{i} F^i(t)(1 - F(t))^{n-i} \\ &- \sum_{i=r}^{n} \binom{n}{i} F^i(t-1)(1 - F(t-1))^{n-i}. \end{aligned}$$

DOI: 10.1201/9781003160632-6

If $F(t^-) = P\{T_i < t\}, i = 1, ..., n$, then (Arnold et al. 2008)

$$P\{T_{r:n} = t\} = \sum_{i=r}^{n} \binom{n}{i} \left\{F^i(t)(1-F(t))^{n-i} - F^i(t^-)(1-F(t^-))^{n-i}\right\}.$$

In particular, the probability mass functions of the extremes $T_{1:n}$ and $T_{n:n}$ can be computed from

$$\begin{aligned}P\{T_{1:n} = t\} &= P\{T_{1:n} > t-1\} - P\{T_{1:n} > t\} \\ &= P\{T_1 > t-1, ..., T_n > t-1\} \\ &\quad - P\{T_1 > t, ..., T_n > t\} \\ &= [1-F(t-1)]^n - [1-F(t)]^n,\end{aligned}$$

and

$$\begin{aligned}P\{T_{n:n} = t\} &= P\{T_{n:n} \le t\} - P\{T_{n:n} \le t-1\} \\ &= P\{T_1 \le t, ..., T_n \le t\} \\ &\quad - P\{T_1 \le t-1, ..., T_n \le t-1\} \\ &= F^n(t) - F^n(t-1).\end{aligned}$$

If the random variables $T_1, ..., T_n$ have supports on non-negative integers, then the moments of $T_{r:n}$ can be computed from

$$E(T_{r:n}^m) = \sum_{i=0}^{m-1} \binom{m}{i} \sum_{j=0}^{\infty} j^i P\{T_{r:n} > j\},$$

for $r = 1, ..., n$, and $m = 1, 2, ...$ (see, e.g. Davies and Dembinska 2018).

Example 6.1. For a system consisting of n independent components, let T_i be the lifetime of the ith component and

$$F(t) = P\{T_i \le t\} = 1 - (1-p)^t,$$

for $i = 1, ..., n$; $t = 1, 2, ...$ If the system has a series structure, then the probability mass function of the lifetime of the system is

$$\begin{aligned}P\{T_{1:n} = t\} &= (1-p)^{n(t-1)} - (1-p)^{nt} \\ &= [1-(1-p)^n]\,[(1-p)^n]^{t-1},\end{aligned}$$

for $t = 1, 2, ...$ That is, the system's lifetime has also a geometric distribution with mean

$$E(T_{1:n}) = \frac{1}{1-(1-p)^n}.$$

On the other hand, if the system has a parallel structure, then the probability mass function of the system's lifetime is

$$P\{T_{n:n} = t\} = \left[1-(1-p)^t\right]^n - \left[1-(1-p)^{t-1}\right]^n.$$

By using the Binomial expansion for the terms $\left[1-(1-p)^t\right]^n$ and $\left[1-(1-p)^{t-1}\right]^n$, we can obtain

$$P\{T_{n:n}=t\}=\sum_{i=1}^{n}(-1)^{i+1}\binom{n}{i}\left[1-(1-p)^i\right]\left[(1-p)^i\right]^{t-1}$$
$$=\sum_{i=1}^{n}(-1)^{i+1}\binom{n}{i}P\{T_{1:i}=t\}. \tag{6.1}$$

The equation (6.1) implies that the lifetime distribution of the parallel system with n components can be written as a generalized mixture of the lifetime distributions of series systems. Indeed,

$$P\{T_{n:n}=t\}=\sum_{i=1}^{n}\omega_i P\{T_{1:i}=t\},$$

where $\omega_i=(-1)^{i+1}\binom{n}{i}$ and $\sum_{i=1}^{n}\omega_i=1$. Thus, the mean time to failure of the parallel system can be computed from

$$E(T_{n:n})=\sum_{i=1}^{n}\omega_i\frac{1}{1-(1-p)^i}.$$

(Eisenberg 2008) has obtained the following lower and upper bounds for the expected value $E(T_{n:n})$:

$$\frac{1}{\lambda}\sum_{k=1}^{n}\frac{1}{k}\leq E(T_{n:n})<1+\frac{1}{\lambda}\sum_{k=1}^{n}\frac{1}{k},$$

where $\lambda=-\ln(1-p)$. ■

If the random variables $T_1,...,T_n$ are independent and

$$P\{T_i=t\}=p_i(1-p_i)^{t-1},$$

for $t=1,2,...,$ then

$$P\{T_{1:n}=t\}=\left(\prod_{i=1}^{n}q_i\right)^{t-1}\left(1-\prod_{i=1}^{n}q_i\right),$$

where $q_i=1-p_i$ (Margolin and Winokur 1967). That is, $T_{1:n}=\min(T_1,...,T_n)$ has also a geometric distribution with mean

$$E(T_{1:n})=\frac{1}{1-\prod_{i=1}^{n}q_i}.$$

Example 6.2. Consider a system consisting of n independent and identical components whose lifetimes are denoted by $T_1,...,T_n$. The system functions if at least k of n components function properly. Assume that the system is equipped with a cold standby component. When the system fails, i.e. at the time when the $(n-k+1)$th failure occurs, a standby component with lifetime Z is put into operation. Thus the lifetime of the system can be represented as

$$T = T_{n-k+1:n} + \min(T_{n-k+2:n} - T_{n-k+1:n}, Z),$$

for $k = 2,...,n$, and $T = T_{n:n} + Z$ when $k = 1$. Let the components have discrete lifetime distribution with

$$F(t) = P\{T_i \leq t\} \ , \ F(t^-) = P\{T_i < t\},$$
$$p(t) = P\{T_i = t\} \ , \ \bar{G}(t) = P\{Z > t\},$$

$i = 1,...,n$. Then, (Dembinska et al. 2021) have shown that

$$P\{T > t\} = n!\sum_{v=0}^{n-k} \frac{1}{v!}\left[\frac{(1-F(t))^{k-1}}{(n-k+1-v)!(k-1)!}\sum_{u=0}^{t}\left(F(u^-)\right)^v \right.$$
$$\left. \times (p(u))^{n-k+1-v}\bar{G}(t-u) + \frac{1}{(n-v)!}(F(t))^v(1-F(t))^{n-v}\right].$$

In the case when the components have continuous lifetime distributions, the same system has been studied by (Eryilmaz 2012a). ■

It is well known that order statistics corresponding to independent identically distributed random variables with continuous distribution exhibit Markovian property. However, Markovian property does not hold, in general, for the discrete case. For a sequence of independent and identically distributed random variables $T_1,...,T_n$ having discrete distribution, define

$$\tilde{T}_i = \#\{j \leq i : T_{j:n} = T_{i:n}\}.$$

Then, as shown by (Rüschendorf 1985) and (Nagaraja 1986), the bivariate sequence $(T_{i:n}, \tilde{T}_i)_{i=1}^n$ forms a Markov chain.

6.2 SYSTEM SIGNATURE

For a binary system consisting of n components, let x_i denote state of the ith component such that $x_i = 1$ if the ith component is functioning and $x_i = 0$ if the ith component has failed, $i = 1,...,n$. Define

$$\phi(x_1,...,x_n) = \begin{cases} 1, & \text{if the system is functioning} \\ 0, & \text{if the system has failed.} \end{cases}$$

The function $\phi(x_1,...,x_n)$ is called the structure function. A system with structure function $\phi(x_1,...,x_n)$ is coherent if it is nondecreasing in each argument, and each component i is relevant to the performance of the system, that is,

$$\phi(x_1,...,x_{i-1},0,x_{i+1},...,x_n) = 0,$$

and

$$\phi(x_1,...,x_{i-1},1,x_{i+1},...,x_n) = 1$$

for some states $x_1,...,x_{i-1},x_{i+1},...,x_n$ of other components $1,...,i-1,i+1,...,n$.

For a coherent system with structure function $\phi : \{0,1\}^n \longrightarrow \{0,1\}$, a path set is a set $P \subseteq \{1,...,n\}$ of its components such that $\phi(x_1,...,x_n) = 1$ if $x_i = 1$ for $i \in P$. That is, a path set is a set of components whose simultaneous functioning puts the system in functioning state.

Example 6.3. For the oil transportation system in Example 2.11, recall that the system consists of n linearly ordered components (pump stations), and it fails if there are at least k consecutively failed components. Assume that each pump station and the system can be in one of two possible states at any time. Then, the structure function of the system is

$$\phi(x_1,...,x_n) = \min(\max(x_1,...,x_k),\max(x_2,...,x_{k+1}),...,\max(x_{n-k+1},...,x_n)),$$

where $x_i = 1$ if the ith pump station is working properly and $x_i = 0$ if the pump station has failed, $i = 1,...,n$. ■

One of the main problems in reliability theory is the computation of system reliability based on component reliabilities. The reliability is simply the probability that the system works under stated conditions. Thus, generally speaking, the reliability of the overall system depends on the physical connection between its components and individual reliabilities of the components. For a binary coherent system that consists of n components with reliabilities $p_1,...,p_n$, the reliability of the overall system is given by

$$R(p_1,...,p_n) = P\{\phi(X_1,...,X_n) = 1\},$$

where $p_i = P\{X_i = 1\}$ is the reliability of the ith component, $i = 1,...,n$. The reliability p_i should be understood as the reliability of the ith component for a fixed period of time. It becomes independent of time variable t when the time is fixed. However, the reliability is in fact a function of time. If the lifetime of the ith component is denoted by T_i, then the time dependent reliability function of the system can be represented by

$$R(\bar{F}_1(t),...,\bar{F}_n(t)),$$

where $\bar{F}_i(t) = P\{T_i > t\}$ is the continuous reliability function of the ith component, $i = 1,...,n$. Thus, to obtain the time dependent reliability function, we simply replace the values of $p_1,...,p_n$ respectively by $\bar{F}_1(t),...,\bar{F}_n(t)$.

For a coherent system consisting of n independent and identical components, the reliability of the system can be computed by conditioning on the total number of working components within the system. Indeed, if p denotes the common reliability value of all components, then the reliability of the system can be computed from

$$R(p) = \sum_{i=1}^{n} r_i(n)p^i(1-p)^{n-i}, \tag{6.2}$$

where $r_i(n)$ is the number of path sets of the system with exactly i components. The calculation of $r_i(n), i = 1,...,n$ is a well-defined combinatorial problem.

Example 6.4. Consider the system with the structure function

$$\phi(x_1, x_2, x_3) = \min(x_1, \max(x_2, x_3)).$$

The system has three path sets which are given by $\{1,2\}, \{1,3\}$, and $\{1,2,3\}$. Thus, we have

$$r_1(n) = 0, r_2(n) = 2, r_3(n) = 1.$$

Therefore, if the system consists of independent and identical components, then from (6.2), its reliability is found to be

$$R(p) = 2p^2(1-p) + p^3. \quad \blacksquare$$

The reliability equation (6.2) can be written as

$$R(p) = \sum_{i=1}^{n} \frac{r_i(n)}{\binom{n}{i}} P\{S_n = i\}, \tag{6.3}$$

where S_n is the total number of working components within the system. By rewriting $P\{S_n = i\}$ in (6.3) as

$$P\{S_n = i\} = P\{S_n \geq i\} - P\{S_n \geq i+1\},$$

we obtain

$$R(p) = \sum_{i=1}^{n} \left[\frac{r_{n-i+1}(n)}{\binom{n}{n-i+1}} - \frac{r_{n-i}(n)}{\binom{n}{n-i}} \right] P\{S_n \geq n-i+1\},$$

where $r_0(n) = 0$. The probability $P\{S_n \geq n-i+1\}$ corresponds to the reliability of the i-out-of-n:F system. The i-out-of-n:F system is a system that consists of n components and fails if at least i components fail. Thus, for a binary coherent system with n independent and identical components, we have

$$R(p) = \sum_{i=1}^{n} s_i P\{S_n \geq n-i+1\},$$

where

$$s_i = \frac{r_{n-i+1}(n)}{\binom{n}{n-i+1}} - \frac{r_{n-i}(n)}{\binom{n}{n-i}},$$

and

$$P\{S_n \geq n-i+1\} = \sum_{j=n-i+1}^{n} \binom{n}{j} p^j (1-p)^{n-j},$$

for $i = 1, \ldots, n$. The vector $\mathbf{s} = (s_1, \ldots, s_n)$ is called signature of the system and it satisfies $\sum_{i=1}^{n} s_i = 1$ (Samaniego 1985, Samaniego 2007). Thus, for a given binary coherent structure, the vector $\mathbf{s} = (s_1, \ldots, s_n)$ is in fact a discrete probability distribution. It should be noted that the signature vector $\mathbf{s} = (s_1, \ldots, s_n)$ only depends on the structure of the system. For a system consisting of independent and identical components having absolutely continuous lifetime distributions, the ith element s_i of the signature vector represents the probability that the system fails upon the occurrence of the ith component failure. The following equation is immediate

$$r_i(n) = \binom{n}{i} \sum_{j=n-i+1}^{n} s_j.$$

For an extensive review and bibliometric analysis of the concept of signature, see, e.g. (Naqvi et al. 2021).

Example 6.5. For the system in Example 6.4, the signature vector is found to be

$$\mathbf{s} = (\frac{1}{3}, \frac{2}{3}, 0).$$

Example 6.6. (Cui and Xie 2005) introduced a generalized k-out-of-n system that is denoted by $((n_1,...,n_N), f, k)$. The system consists of N modules ordered in a line and the ith module is composed of n_i components in parallel, where $n_i \geq 1, i = 1,2,...,N$. The $((n_1,...,n_N), f, k)$ system fails iff there exist at least f failed components, or at least k consecutive failed modules. Let $r_i(n,N,f,k)$ denote the number of path sets of size i of the $((n_1,...,n_N), f, k)$ system. Then, (Eryilmaz and Tuncel 2015) obtained the following formula for $r_i(n,N,f,k)$:

$$r_i(n,N,f,k) = \sum_{j=1}^{N} M(j,k,N) S(j,i),$$

where

$$M(j,k,N) = \sum_{i=0}^{j+1} (-1)^i \binom{j+1}{i} \binom{N-ik}{j},$$

and

$$S(j,i) = \sum_{a=1}^{i-j+1} \binom{n}{a} S(j-1, i-a),$$

for $1 < j \leq i$, $S(1,i) = \binom{n}{i}$ for $i \geq 1$ when $n \cdot N - f + 1 \leq i \leq j \cdot n$. Thus, if the components within the system are independent and identical with common reliability value p, then the reliability of the entire system can be computed from

$$\sum_{i=1}^{n \cdot N} r_i(n,N,f,k) p^i (1-p)^{n \cdot N - i}.$$

Example 6.7. The m-consecutive-k-out-of-n:F system with overlapping runs is a system consisting of n linearly ordered components and fails if and only if there are at least m overlapping runs of k consecutive failures (Agarwal and Mohan 2008). For $m = 1$, the system is same with the linear consecutive-k-out-of-n:F system. (Eryilmaz 2012b) obtained the following combinatorial equation to compute the number of path sets of the m-consecutive-k-out-of-n:F system with overlapping runs:

$$\begin{aligned} r_i(n) &= N(i+1, k+1, n+1) \\ &+ \sum_{x=1}^{m-1} \sum_{a=1}^{\min(i+1,x)} \binom{i+1}{a} \binom{x-1}{a-1} N(i-a+1, k+1, n-x-ak+1), \end{aligned}$$

for $1+\left[\frac{n-m-k+1}{k}\right] \leq i < n$ and $r_n(n)=1$, where

$$N(a,b,c) = \sum_{j=0}^{\min(a,\left[\frac{c-a}{b-1}\right])} (-1)^j \binom{a}{j} \binom{c-j(b-1)-1}{a-1},$$

and $N(0,b,0)=1$ for $b>0$. Thus, if the system consists of independent and identical component with common reliability value p, then the reliability of the system can be computed from

$$R = \sum_{i=1+\left[\frac{n-m-k+1}{k}\right]}^{n} r_i(n) p^i (1-p)^{n-i}.$$

Further, the ith element of the signature vector can be obtained from

$$s_i = \frac{r_{n-i+1}(n)}{\binom{n}{n-i+1}} - \frac{r_{n-i}(n)}{\binom{n}{n-i}}.$$

6.3 POWER SERIES CLASS OF DISTRIBUTIONS

A random variable T is said to have a power series distribution if its probability mass function has the form

$$P\{T=t\} = \frac{a_t \theta^t}{b(\theta)}, \; t=1,2,..., \tag{6.4}$$

where $a_t \geq 0$ depends on only t, $b(\theta)=\sum_{t=1}^{\infty} a_t \theta^t$ and $\theta \in (0,s)$ (s can be ∞) is such that $b(\theta)$ is finite. The mean of the random variable T is

$$E(T) = \theta \frac{b'(\theta)}{b(\theta)}.$$

For the details of power series class of distributions see, e.g. (Noack 1950). The power series class of distributions (6.4) reduces to the well-known discrete distributions for particular selections of a_t and $b(\theta)$. In Table 6.1, we list useful quantities for some power series distributions.

Table 6.1
Useful Quantities for Some Power Series Distributions

Distribution	a_t	$b(\theta)$	$b'(\theta)$	$b''(\theta)$	s
Modified Poisson	$1/t!$	$e^\theta - 1$	e^θ	e^θ	∞
Geometric	1	$\theta(1-\theta)^{-1}$	$(1-\theta)^{-2}$	$2(1-\theta)^{-3}$	1
Logarithmic	$1/t$	$-\ln(1-\theta)$	$(1-\theta)^{-1}$	$(1-\theta)^{-2}$	1

The distribution given by (6.4) has been used to obtain new distributions. For example, (Chahkandi and Ganjali 2009), (Lu and Chiang 2018) studied the distribution of the random minima $\min(Y_1,...,Y_T)$ by assuming (6.4) for the distribution of the random variable T. (Choudhury et al. 2019) studied stress-strength reliability when the stress and strength distributions are different members of the power series class of distributions.

Example 6.8. Let T_1 represent the applied stress on a unit by the operating environment and T_2 be the strength or resistance of the unit, then $R = P\{T_1 \leq T_2\}$ is the probability that the unit performs satisfactorily, i.e. the reliability of the unit. Assume that

$$P\{T_1 = t\} = \frac{a_t \theta^t}{b(\theta)}, \ t = 1,2,...,$$

and

$$P\{T_2 = t\} = \frac{c_t \beta^t}{d(\beta)}, \ t = 1,2,....$$

Then, under the assumption that T_1 and T_2 are independent, (Choudhury et al. 2021) have shown that

$$R = P\{T_1 \leq T_2\} = \sum_{t_2=1}^{\infty} P\{T_1 \leq t_2\} P\{T_2 = t_2\}$$

$$= \frac{\sum_{t_2=1}^{\infty} \left\{ \sum_{t_1=1}^{t_2} a_{t_1} \theta^{t_1} \right\} c_{t_2} \beta^{t_2}}{b(\theta) d(\beta)}.$$

Example 6.9. Consider a system consisting of n independent components. Let X_i denote the lifetime of the ith component and $F(x) = P\{X_i \leq x\}, i \geq 1$. Assume that the random variable T is independent of $X_1, X_2, ...$ and follows the distribution given by (6.4). Then, the random variable $\max(X_1,...,X_T)$ represents the lifetime of a parallel system with random number of components. (Eryilmaz 2017) has shown that

$$\begin{aligned} F^*(x) &= P\{\max(X_1,...,X_T) \leq x\} \\ &= \sum_{t=1}^{\infty} F^t(x) P\{T = t\} \\ &= \frac{1}{b(\theta)} \sum_{t=1}^{\infty} a_t (\theta F(x))^t \\ &= \frac{b(\theta F(x))}{b(\theta)}. \end{aligned}$$

6.4 PANJER'S FAMILY OF DISTRIBUTIONS

The random variable T whose probability mass function satisfies the following recursive relation is said to be a member of the Panjer's family of distributions.

$$P\{T = t\} = \left(a + \frac{b}{t}\right) P\{T = t - 1\}, \tag{6.5}$$

for $t = 1,2,...$ (Panjer 1981). The well-known Binomial, Poisson, Geometric, and Negative Binomial distributions are the members of this family. Indeed, if $a = 0, b = \lambda, P\{T = 0\} = e^{-\lambda}$, then the Poisson distribution is obtained. For

$$a = -\frac{p}{1-p}, \; b = (n+1)\frac{p}{1-p},$$

and $P\{T = 0\} = (1-p)^n$, (6.5) gives a recursion for the Binomial distribution.

Consider the compound random variable $S = \sum_{i=1}^{T} Y_i$ when T obeys (6.5). Let $Y_1, Y_2, ...$ be a sequence of positive valued independent and identically distributed random variables with common probability mass function $f_Y(y) = P\{Y_i = y\}$, independent of T. Then,

$$P\{S = t\} = \sum_{u=1}^{t} P\{S = t-u\} P\{Y = u\} \left(a + b\frac{u}{t}\right),$$

for $t = 1,2,...$ (Panjer 1981).

Important generalizations of (6.5) have been presented by (Sundt 1992).

Example 6.10. Let T have Poisson distribution with probability mass function

$$P\{T = t\} = \frac{e^{-\lambda}\lambda^t}{t!}, \quad t = 0,1,2,...,$$

and

$$P\{Y_i = y\} = p(1-p)^{y-1}, \quad y = 1,2,....$$

Then, the probability mass function of $S = \sum_{i=1}^{T} Y_i$ can be recursively computed from

$$P\{S = t\} = \frac{\lambda p}{t} \sum_{u=1}^{t} u P\{S = t-u\} (1-p)^{u-1},$$

with $P\{S = 0\} = e^{-\lambda}$.

REFERENCES

1. Agarwal, M. and Mohan, P. 2008. GERT analysis of m-consecutive-k-out-of-n:F system with overlapping runs and $(k-1)$-step markov dependence. *International Journal of Operational Research*. 3: 36-51.
2. Arnold, B.C., Balakrishnan, N. and Nagaraja, H.N. *A first course in order statistics*. Classics in Applied Mathematics. Society for Industrial and Applied Mathematics.
3. Chahkandi, M. and Ganjali, M. 2009. On some lifetime distributions with decreasing failure rate. *Computational Statistics and Data Analysis* 53: 4433-4440.
4. Choudhury, M.M., Bhattacharya, R. and Maiti, S.S. 2021. On estimating reliability function for the family of power series distribution. *Communications in Statistics-Theory and Methods* 50: 2801-2830.
5. Cui, L. and Xie, M. 2005. On a generalized k-out-of-n system and its reliability. *International Journal of Systems Science* 36: 267-274.
6. David, H.A. and Nagaraja, H.N. 2003. *Order statistics*. John Wiley & Sons, New Jersey.

7. Davies, K. and Dembinska, A. 2018. Computing moments of discrete order statistics from non-identical distributions. *Journal of Computational and Applied Mathematics* 328: 340-354.
8. Dembinska, A., Nikolov, N.I. and Stoimenova, E. 2021. Reliability properties of k-out-of-n systems with one cold standby unit. *Journal of Computational and Applied Mathematics* 388: 113289.
9. Eisenberg, B. 2008. On the expectation of the maximum of IID geometric random variables. *Statistics and Probability Letters* 78: 135-143.
10. Eryilmaz, S. 2012a. On the mean residual life of a k-out-of-n: G system with a single cold standby component. *European Journal of Operational Research* 222: 273-277.
11. Eryilmaz, S. 2012b. m-consecutive-k-out-of-n:F system with overlapping runs: signature-based reliability analysis. *International Journal of Operational Research* 15: 64-73.
12. Eryilmaz, S. 2017. A note on optimization problems of a parallel system with a random number of units. *International Journal of Reliability, Quality and Safety Engineering* 24: 1750022.
13. Eryilmaz, S. and Tuncel, A. 2015. Computing the signature of a generalized k-out-of-n system. *IEEE Transactions on Reliability* 64: 766-771.
14. Lu, W. and Chiang, J-Y. 2018. On some life distributions with flexible failure rate. *Quality Technology and Quantitative Management* 15: 413-433.
15. Margolin, B.H. and Winokur, H.S. 1967. Exact moments of the order statistics of the geometric distribution and their relation to inverse sampling and reliability of redundant systems. *Journal of the Americal Statistical Association* 62: 915-925.
16. Nagaraja, H.N. 1986. A note on conditional Markov property of discrete order statistics. *Journal of Statistical Planning and Inference* 13; 37-43.
17. Naqvi, S., Chan, P.S. and Mishra, D.B. 2021. System signatures: A review and bibliometric analysis. *Communications in Statistics - Theory and Methods*, DOI: 10.1080/03610926.2021.1937653.
18. Noack, A. 1950. A class of random variables with discrete distributions. *Annals of Mathematical Statistics* 21: 127-132.
19. Panjer, H. H. 1981. Recursive evaluation of a family of compound distributions. *ASTIN Bulletin* 12: 22-26.
20. Rüschendorf, L. 1985. Two remarks on order statistics. *Journal of Statistical Planning and Inference* 11, 71-74.
21. Samaniego, F. J. 1985. On closure of the IFR class under formation of coherent systems. *IEEE Transactions on Reliability* 34: 69-72.
22. Samaniego, F. J. 2007. *System signatures and their applications in engineering reliability*. Springer, New York.
23. Sundt, B., 1992. On some extensions of Panjer's class of counting distributions. *ASTIN Bulletin* 22: 61-80.

7 Other Discrete Stochastic Models Involving Dependency

Dependence is an important concept in stochastic modeling. This section is devoted to discrete stochastic structures for modeling dependency. First, the concept of exchangeable dependence is presented. Then, the particular model which is known to be previous-sum dependent model is mentioned. Finally, the distribution of the number of successes in a sequence of dependent binary trials is considered.

7.1 EXCHANGEABLE DEPENDENCE

The exchangeable dependence is of special importance in statistics and probability. Exchangeable random variables are also known to be symmetrically dependent and appear in many stochastic problems involving urn models, biological models, and reliability models. See, e.g. (Hill et al. 1987), (Lau 1992), (George and Bowman 1995), (Navarro and Rychlik 2007), (Mahmoud 2008), (Navarro et al. 2010). The models based on exchangeable binary random variables have been studied by (Eryilmaz and Demir 2007), (Yu and Zelterman 2008), (Eryilmaz 2010), (Zaigraev and Kaniovski 2010), and (Banerjee and Richardson 2013).

Let X_1 and X_2 be two random variables that may take two possible values either 0 or 1. If the random variables X_1 and X_2 are exchangeable dependent, then we have

$$P\{X_1 = i\} = P\{X_2 = i\} \text{ for } i = 0, 1,$$

and

$$P\{X_1 = 1, X_2 = 0\} = P\{X_1 = 0, X_2 = 1\}.$$

Example 7.1. Consider a system with two exchangeable components. Let X_i denote the state of the ith component, $i = 1, 2$. Assume that

$$P\{X_1 = 1, X_2 = 0\} = P\{X_1 = 0, X_2 = 1\} = 0.25,$$

$$P\{X_1 = 1, X_2 = 1\} = 0.4.$$

If the system has a parallel structure, then the probability that the system works is

$$\begin{aligned} &P\{X_1 = 1, X_2 = 0\} + P\{X_1 = 0, X_2 = 1\} + P\{X_1 = 1, X_2 = 1\} \\ &= 0.25 + 0.25 + 0.4 \\ &= 0.9. \quad \blacksquare \end{aligned}$$

DOI: 10.1201/9781003160632-7

In general, a finite sequence of random variables $X_1, X_2, ..., X_n$ is said to be exchangeable if for each n and any permutation $\pi(1), ..., \pi(n)$ of $1, ..., n$,

$$P\{X_1 \leq x_1, ..., X_n \leq x_n\} = P\{X_{\pi(1)} \leq x_1, ..., X_{\pi(n)} \leq x_n\}. \tag{7.1}$$

The exchangeability implies that all subsets of $X_1, ..., X_n$ with the same size has the same joint distribution.

Example 7.2. Let X_1 and X_2 denote the lifetimes of two dependent units. The joint cumulative distribution function of X_1 and X_2 is given by

$$P\{X_1 \leq x_1, X_2 \leq x_2\} = (1 - e^{-\theta_1 x_1})(1 - e^{-\theta_2 x_2})\left[1 + \alpha e^{-\theta_1 x_1} e^{-\theta_2 x_2}\right],$$

for $-1 \leq \alpha \leq 1$. The random variables X_1 and X_2 are exchangeable if $\theta_1 = \theta_2$. ■

Let $X_1, X_2, ..., X_n$ be exchangeable binary random variables taking values either 0 or 1, and $S_n = \sum_{i=1}^{n} X_i$. (George and Bowman 1995) obtained the following expression for the distribution of S_n :

$$P\{S_n = s\} = \binom{n}{s} \sum_{i=0}^{n-s} (-1)^i \binom{n-s}{i} \lambda_{s+i}, \tag{7.2}$$

where $\lambda_0 = 1$ and $\lambda_k = P\{X_1 = 1, ..., X_k = 1\}, k = 1, 2, ...$.

Various models is obtained as a special case of (7.2). For $\lambda_k = p^k, k = 0, 1, ...$ in (7.2), the random variable S_n has a binomial distribution with parameters n and p. The random variable S_n has a beta-binomial distribution when

$$\lambda_k = \frac{B(a+k, b)}{B(a, b)},$$

for $k = 0, 1, ...$ where $B(a, b) = \frac{\Gamma(a)\Gamma(b)}{\Gamma(a+b)}$ denotes Beta function and $\Gamma(.)$ is the Gamma function.

Example 7.3. Consider a system consisting of n independent components. Suppose that the components of this system are subject to a common random stress Z. Denote by $Y_1, Y_2, ..., Y_n$ the random strengths of the components. The component fails if the stress exceeds the components' strength at any moment. That is, if $Y_i > Z$ then the ith component functions otherwise fails, $i = 1, 2, ..., n$. Let's define

$$X_i = \begin{cases} 1, & \text{if } Y_i > Z \\ 0, & \text{if } Y_i \leq Z, \end{cases}$$

$i = 1, 2, ..., n$. The random variables $X_1, X_2, ..., X_n$ are exchangeable if the random variables $Y_1, Y_2, ..., Y_n$ have the common distribution. Let $F(y) = P\{Y_i \leq y\}, i = 1, 2, ..., n$ and $G(z) = P\{Z \leq z\}$. If the cumulative distribution functions F and G are continuous, then

$$\begin{aligned} \lambda_k &= P\{X_1 = 1, ..., X_k = 1\} \\ &= P\{Y_1 > Z, ..., Y_k > Z\} \\ &= \int_0^{\infty} (1 - F(z))^k dG(z). \end{aligned}$$

Assume that the entire system functions iff at least k components function properly. Then, $P\{S_n \geq k\}$ gives the working probability (reliability) of the overall system, where $S_n = \sum_{i=1}^{n} X_i$. Let

$$F(y) = 1 - e^{-\lambda y}\,, y \geq 0,$$

and

$$G(z) = 1 - e^{-\theta z}\,, z \geq 0.$$

Then,

$$\begin{aligned} \lambda_k &= \int_0^\infty (1 - F(z))^k dG(z) \\ &= \int_0^\infty e^{-k\lambda z} \theta e^{-\theta z} dz \\ &= \frac{\theta}{\theta + k\lambda}. \end{aligned}$$

Thus, the reliability of the k-out-of-n system under this particular stress-strength setup is

$$\begin{aligned} P\{S_n \geq k\} &= \sum_{s=k}^{n} \binom{n}{s} \sum_{i=0}^{n-s} (-1)^i \binom{n-s}{i} \frac{\theta}{\theta + (s+i)\lambda} \\ &= \sum_{s=k}^{n} \binom{n}{s} \sum_{i=0}^{n-s} (-1)^i \binom{n-s}{i} \frac{1}{1 + (s+i)\frac{\lambda}{\theta}}. \end{aligned}$$

The reliability of the system can be easily computed for given n, k, θ, and λ. See, e.g. (Bhattacharyya and Johnson 1974) for further discussion on stress-strength reliability in multicomponent setup. ■

Next, consider a weak exchangeable binary sequence of order m which consists of m random vectors such that the elements within each random vector are exchangeable dependent in the usual sense and the different random vectors have different distributions. For $m = 1$, a weak exchangeable sequence of order m corresponds to usual exchangeable sequence. Statistically, for $n = n_1 + n_2 + ... + n_m$, a finite sequence of binary random variables

$$(X_1, X_2, ..., X_n) = (X_{i_1}^{(1)}, ..., X_{i_{n_1}}^{(1)}, X_{i_{n_1+1}}^{(2)}, ..., X_{i_{n_1+n_2}}^{(2)}, ..., X_{i_{n-n_m+1}}^{(m)}, ..., X_{i_n}^{(m)})$$

is called weak exchangeable sequence of order m ($m \geq 1$) if for $(i_1, i_2, ..., i_n) \in \{1, 2, ..., n\}$,

$$P\left\{X_{i_1}^{(1)} = x_1, ..., X_{i_{n_1}}^{(1)} = x_{n_1}, X_{i_{n_1+1}}^{(2)} = x_{n_1+1}, ..., X_{i_{n_1+n_2}}^{(2)} = x_{n_1+n_2}, ..., X_{i_{n-n_m+1}}^{(m)} = x_{n-n_m+1}, ..., X_{i_n}^{(m)} = x_n\right\}$$
$$= P\left\{X_{\pi_1(i_1)}^{(1)} = x_1, ..., X_{\pi_1(i_{n_1})}^{(1)} = x_{n_1}, X_{\pi_2(i_{n_1+1})}^{(2)} = x_{n_1+1}, ..., X_{\pi_2(i_{n_1+n_2})}^{(2)} = x_{n_1+n_2}, ..., X_{\pi_m(i_{n-n_m+1})}^{(m)} = x_{n-n_m+1}, ..., X_{\pi_m(i_n)}^{(m)} = x_n\right\}$$

for any permutations $(\pi_1(i_1), ..., \pi_1(i_{n_1}))$ of $\{i_1, ..., i_{n_1}\}$, $(\pi_2(i_{n_1+1}), ..., \pi_2(i_{n_1+n_2}))$ of $\{i_{n_1+1}, ..., i_{n_1+n_2}\}$, ..., $(\pi_m(i_{n-n_m+1}), ..., \pi_m(i_n))$ of $\{i_{n-n_m+1}, ..., i_n\}$, $x_1, ..., x_n \in \{0, 1\}$ (Eryilmaz 2017).

Let $S_{n_i}^{(i)}$ denote the sum of the n_i exchangeable binary random variables of type i, $i = 1, 2, ..., m$. (Eryilmaz 2017) has obtained the following expression for the joint distribution of the random variables $S_{n_1}^{(1)}, ..., S_{n_m}^{(m)}$:

$$P\left\{S_{n_1}^{(1)} = s_1, S_{n_2}^{(2)} = s_2, ..., S_{n_m}^{(m)} = s_m\right\}$$
$$= \binom{n_1}{s_1} ... \binom{n_m}{s_m}$$
$$\times \sum_{i_1=0}^{n_1-s_1} \cdots \sum_{i_m=0}^{n_m-s_m} (-1)^{i_1+...+i_m} \binom{n_1 - s_1}{i_1} ... \binom{n_m - s_m}{i_m} \lambda_{s_1+i_1,...,s_m+i_m},$$

where

$$\lambda_{s_1,...,s_m} = P\left\{X_1^{(1)} = 1, ..., X_{s_1}^{(1)} = 1, ..., X_1^{(m)} = 1, ..., X_{s_m}^{(m)} = 1\right\},$$

for $s_i = 0, 1, 2, ...$, $i = 1, 2, ..., m$ with $\lambda_{0,...,0} = 1$.

An infinite sequence of binary random variables $X_1, X_2, ...$ is infinitely exchangeable if its every finite subsequence is exchangeable. According to the fundamental theorem of de Finetti, there is a random variable Θ supported on $(0, 1)$ with cumulative distribution function $F(p) = P\{\Theta \leq p\}$ such that

$$P\{X_1 = x_1, ..., X_n = x_n\} = \int_0^1 p^m (1 - p)^{n-m} dF(p), \tag{7.3}$$

where $m = \sum_{i=1}^n x_i$.

From de Finetti's Theorem, one can conclude that infinitely exchangeable random variables are conditionally independent and identically distributed random variables given $\Theta = p$. Therefore, if S_n denotes the total number of successes in the first n trials of an infinite exchangeable binary sequence, then

$$P\{S_n \leq i\} = \int_0^1 P\{S_n^*(p) \leq i\} dF(p), \tag{7.4}$$

where the random variable $S_n^*(p)$ has a Binomial distribution with parameters n and p, i.e.

$$P\{S_n^*(p)=m\}=\binom{n}{m}p^m(1-p)^{n-m},$$

$m=0,1,...,n$.

Start-up demonstration tests considered in Examples 2.10 and 4.5 have been examined by (Eryilmaz and Chakraborti 2008), (Eryilmaz 2008), (Rakitzis and Antzoulakos 2013), and (Balakrishnan et al. 2014) when the individual start-ups form a binary sequence of exchangeable trials.

A more general dependence model which is known to be partial exchangeability is defined by the help of Markov chains. A sequence $X_1,X_2,...$ of binary random variables taking values on $I=\{0,1\}$ is said to be partially exchangeable, or Markov exchangeable if

$$P\{X_1=\sigma_1,...,X_n=\sigma_n\}=P\{X_1=\tau_1,...,X_n=\tau_n\}. \tag{7.5}$$

for two sequences $\sigma,\tau\in I^n$, which have the same starting state and the same transition counts.

A process $\{X_i\}_{i\geq 1}$ is said to be recurrent if

$$P\{X_n=X_1 \text{ for infinitely many } n\}=1.$$

(Diaconis and Freedman 1980) proved that $\{X_i\}_{i\geq 1}$ is a mixture of Markov chains if and only if $\{X_i\}_{i\geq 1}$ is partially exchangeable for recurrent processes. More explicitly, let $\{X_i\}_{i\geq 1}$ be a Markov chain with unknown stationary transition matrix

$$P=\begin{bmatrix} p_{00} & p_{01} \\ p_{10} & p_{11} \end{bmatrix}.$$

If $t_{ij}(n)$ denotes the number of transitions from state i to state j in the sequence $x_1,x_2,...,x_n$ then with fixed x_1

$$\begin{aligned} &P\{X_1=x_1,...,X_n=x_n\} \\ &=\int p_{00}^{t_{00}(n)}p_{01}^{t_{01}(n)}p_{10}^{t_{10}(n)}p_{11}^{t_{11}(n)}\mu(dp_{00},dp_{01},dp_{10},dp_{11}) \end{aligned} \tag{7.6}$$

holds for some μ if and only if (7.5) holds, provided P is recurrent. The equation (7.6) can also be written as

$$\begin{aligned} &P\{X_1=x_1,...,X_n=x_n\} \\ &=\int_0^1\int_0^1 p_{00}^{t_{00}(n)}(1-p_{00})^{t_{01}(n)}p_{10}^{t_{10}(n)}(1-p_{10})^{t_{11}(n)}\mu(dp_{00},dp_{10}). \end{aligned}$$

Various existing models for binary sequences can be obtained from partial exchangeability.

* If the mixing measure μ concentrated on the set $\{(p,p) : p \in [0,1]\}$ then the resulting sequence is a sequence of exchangeable binary random variables.
* If μ is a point mass at $(p^\star, p^\star), p^\star \in (0,1)$, then a sequence of independent identically distributed binary random variables is obtained.
* An usual Markov chain is obtained whenever μ is a point mass on an arbitrary point of the unit square $[0,1]^2$.

7.2 PREVIOUS-SUM DEPENDENT MODEL

Let $\{X_i\}_{i\geq 1}$ be a sequence of binary trials satisfying

$$P\{X_n = 1 \mid X_1, X_2, ..., X_{n-1}\} = a_n + b_n S_{n-1}, \tag{7.7}$$

for $n \geq 2$ and a_n and b_n satisfy $0 < a_n < 1$, and $0 < a_n + (n-1)b_n < 1$ for $n \geq 1$, where $S_n, n \geq 1$ denotes the total number of successes in $X_1, X_2, ..., X_n$ and for convenience $P\{S_0 = 0\} = 1$. According to (7.7), the conditional distribution of X_n given $X_1, X_2, ..., X_{n-1}$ depends on the total number of successes in $X_1, X_2, ..., X_{n-1}$. Therefore, the corresponding model is known to be previous-sum dependent model. This model was first considered by (Vellaisamy 1996) to get new generalized Binomial distributions and it has been used for modeling a dependent production process in (Vellaisamy and Sankar 2005). In (7.7), if $a_i = p_i$ and $b_i = 0$, for $i \geq 2$, then the model corresponds to independent, nonidentical Bernoulli trials.

(Vellaisamy and Upadhye 2007) have shown that

$$P\{X_n = 1 | S_{n-1}\} = a_n + b_n S_{n-1}. \tag{7.8}$$

The following recursive scheme also holds true for the distribution of the total number of successes.

$$P\{S_n = x\} = (a_n + b_n(x-1))P\{S_{n-1} = x-1\} + (1 - a_n - b_n x)P\{S_{n-1} = x\}. \tag{7.9}$$

Example 7.4. According to the TSTF (Total Successes Total Failures) type start-up demonstration test, the equipment under test is accepted if k successful start-ups are observed before d failures. For an illustration, let $k = 5$ and $d = 2$, and the outcomes of the individual start-ups are 110110. Then, the equipment is rejected since $d = 2$ failures are obtained before $k = 5$ successes. In this case, the test length (the number of start-ups until either acceptance or rejection) is six. Let $T_k^{(1)}$ denote the number of start-ups until the first k successes and $T_d^{(0)}$ denote the number of start-ups until the first d failures. (Yalcin and Eryilmaz 2012) obtained the following equation for the probability of accepting the equipment under the model (7.7):

$$P\left\{T_k^{(1)} < T_d^{(0)}\right\} = \sum_{i=k}^{k+d-1} [a_i + b_i(k-1)] P\{S_{i-1} = k-1\}. \tag{7.10}$$

If $a_i = p$ and $b_i = 0$ for $i \geq 2$, then (7.10) corresponds to the probability of accepting the equipment when the individual start-ups are independent and identical with the probability of successful start-up p. In other words, (7.10) is the probability of getting k successes before d failures in a sequence of classical Bernoulli trials. In this case, we have

$$P\left\{T_k^{(1)} < T_d^{(0)}\right\} = \sum_{i=k}^{k+d-1} \binom{i-1}{k-1} p^k (1-p)^{i-k}$$
$$= \sum_{j=0}^{d-1} \binom{k+j-1}{k-1} p^k (1-p)^j$$

which coincides with equation (1) in (Smith and Griffith 2008).

7.3 THE DISTRIBUTION OF THE NUMBER OF SUCCESSES

The distribution of the number of successes in a binary sequence consisting of two types of trials as either success or failure has attracted great deal of attention in probability. For an arbitrary sequence $X_1, X_2, ..., X_n$ of binary trials with two possible outcomes "1" or "0", the exact distribution of the number of successes $S_n = \sum_{i=1}^{n} X_i$ is given by (Blom et al. 1994)

$$P\{S_n = k\} = \sum_{j=k}^{n} (-1)^{j-k} \binom{j}{k} T_j,$$

where

$$T_j = \sum_{1 \le i_1 < i_2 < ... < i_j \le n} P\left\{X_{i_1} = 1, X_{i_2} = 1, ..., X_{i_j} = 1\right\}.$$

In particular, if $X_1, X_2, ..., X_n$ are exchangeable dependent, then

$$T_j = \binom{n}{j} P\{X_1 = 1, X_2 = 1, ..., X_j = 1\},$$

and hence

$$P\{S_n = k\} = \sum_{j=k}^{n} (-1)^{j-k} \binom{j}{k} \binom{n}{j} P\{X_1 = 1, X_2 = 1, ..., X_j = 1\},$$

for $k = 0, 1, ..., n$.

(Eryilmaz 2021) has obtained the following expression for the distribution of S_n when $X_1, ..., X_n$ is a weak exchangeable sequence of order m:

$$P\{S_n = s\}$$
$$= \sum_{\mathbf{s}} \prod_{i=1}^{m} \binom{n_i}{s_i} \sum_{i_1=0}^{n_1-s_1} \cdots \sum_{i_m=0}^{n_m-s_m} (-1)^{i_1+...+i_m} \prod_{j=1}^{m} \binom{n_j - s_j}{i_j} \lambda_{s_1+i_1,...,s_m+i_m}, \qquad (7.11)$$

the first sum in (7.11) is taken over all $s_1, s_2, ..., s_m$ such that $s_1 + s_2 + ... + s_m = s$. The equation (7.11) generalizes (7.2) from exchangeable sequence to weak exchangeable sequence of order m.

7.4 *m*-DEPENDENT SEQUENCE OF RANDOM VARIABLES

A sequence of random variables $X_1, X_2, \ldots$ is called m-dependent for a given fixed m if two sets of random variables $\{X_1, \ldots, X_i\}$ and $\{X_j, X_{j+1}, \ldots\}$ are independent whenever $j - i > m$, see, e.g. (Shang 2012). The following Normal approximation was established by (Hoeffding and Robbins 1949) for a stationary sequence of random variables.

Let $X_1, \ldots, X_n, \ldots$ be a stationary sequence of random variables and let it also be m-dependent with m held fixed. Then, if $E|X_1|^3 < \infty$,

$$\frac{\sum_{i=1}^{n}(X_i - E(X_i))}{\sqrt{n}} \to N(0, \sigma^2), \tag{7.12}$$

where

$$\sigma^2 = Var(X_1) + 2\sum_{i=2}^{m+1}(E(X_1 X_i) - E(X_1)E(X_i)).$$

Example 7.5. Let $Y_1, Y_2, \ldots, Y_n$ be a sequence of independent and identically distributed random variables with

$$P\{Y_i = y\} = p(1-p)^{y-1},\ y = 1, 2, \ldots$$

Denote by R_n the total number of runs in $Y_1, Y_2, \ldots, Y_n$. That is,

$$R_n = 1 + \sum_{i=1}^{n-1} X_i,$$

where

$$X_i = \begin{cases} 1, & \text{if } Y_i \neq Y_{i+1}, \\ 0 & \text{otherwise} \end{cases}$$

for $i = 1, 2, \ldots n-1$. Manifestly, $X_1, X_2, \ldots$ is a sequence of 1-dependent stationary random variables. (Grabner et al. 2003) (see also (Eryilmaz 2006)) obtained the following expressions for the expected value and variance of R_n :

$$E(R_n) = \frac{2q}{1+q}n + \frac{p}{1+q},$$

and

$$Var(R_n) = \frac{2q(1-q)^2(2+q^2)}{(1-q^3)(1+q)^2}n - \frac{2q(1-q)^2(3-q+q^2)}{(1-q^3)(1+q)^2},$$

where $q = 1 - p$.

(Eryilmaz 2006) obtained the following result by applying (7.12):

$$P\left\{R_n \le 1 + (n-1)\frac{2q}{1+q} + x\sigma\sqrt{n}\right\} = \Phi(x) + O(n^{-1/2}), \tag{7.13}$$

where

$$\sigma^2 = \frac{2q(2+q^2)(1-q)^2}{(1-q^3)(1+q)^2},$$

and $\Phi(x)$ is the cumulative distribution function of the standard Normal random variable. The result (7.13) was first obtained by (Grabner et al. 2003) using a probability generating function approach.

REFERENCES

1. Bhattacharyya, G.K. and Johnson, R.A. 1974. Estimation of reliability in a multicomponent stress strength model. *Journal of the American Statistical Association* 69: 966 970.
2. Balakrishnan, N., Koutras, M.V. and Milienos, F.S. 2014. Some binary start-up demonstration tests and associated inferential methods. *Annals of the Institute of Statistical Mathematics* 66: 759-787.
3. Banerjee, M. and Richardson, T. 2013. Exchangeable Bernoulli random variables and Bayes' postulate. *Electronic Journal of Statistics* 7: 2193-2208.
4. Blom, G., Holst, L. and Sandell, D. 1994. Problems and snapshots from the world of probability, Springer, New York.
5. Diaconis, P., Freedman, D. 1980. De Finetti's theorem for Markov chains. *The Annals of Probability* 8: 115-130.
6. Eryilmaz, S. 2006. A note on runs of geometrically distributed random variables. *Discrete Mathematics* 306: 1765-1770.
7. Eryilmaz, S. and Demir, S. 2007. Success runs in a sequence of exchangeable binary trials. *Journal of Statistical Planning and Inference* 137: 2954-2963.
8. Eryilmaz, S. 2008. Run statistics defined on the multi-color urn model. *Journal of Applied Probability* 45, 1007-1023.
9. Eryilmaz, S. and Chakraborti, S. 2008. On start-up demonstration tests under exchangeability. *IEEE Transactions on Reliability* 57: 627-632.
10. Eryilmaz, S. 2010. Discrete scan statistics generated by exchangeable binary trials. *Journal of Applied Probability* 47: 1084-1092.
11. Eryilmaz, S. 2017. The concept of weak exchangeability and its applications. *Metrika* 80: 259-271.
12. George, E.O. and Bowman, D. 1995. A full likelihood procedure for analysing exchangeable binary data. *Biometrics* 51: 512-523.
13. Good, P.I. 2002. Extensions of the concept of exchangeability and their applications. *Journal of Modern Applied Statistical Methods*, 1, 243-247.
14. Grabner, P.J., Knopfmacher, A. and Prodinger, H. 2003. Combinatorics of geometrically distributed random variables: run statistics, Theoretical Computer Science 297: 261-270.
15. Hill, B.M., Lane, D. and Sudderth, W. 1987. Exchangeable urn processes. *The Annals of Probability* 15: 1586-1592.
16. Hoeffding, W. and Robbins, H. 1949. The central limit theorem for dependent random variables. *Ann. Math. Statist.* 20: 773-780.
17. Lau T.S. 1992. The reliability of exchangeable binary systems. *Statistics and Probability Letters* 13: 153-158.
18. Mahmoud, H. 2008. Polya urn models, Chapman-Hall, Orlando, Florida.
19. Navarro, J. and Rychlik, T. 2007. Reliability and expectation bounds for coherent systems with exchangeable components. *Journal of Multivariate Analysis* 98: 102-113.

20. Navarro, J., Spizzichino, F. and Balakrishnan, N. 2010. Applications of average and projected systems to the study of coherent systems. *Journal of Multivariate Analysis* 101: 1471-1482.
21. Rakitzis, A.C. and Antzolulakos, D.L. 2013. A start-up demonstration test based on exchangeable binary trials. *IEEE Transactions on Reliability* 62: 736-745.
22. Shang, Y. 2012. A central limit theorem for randomly indexed *m*-dependent random variables. *Filomat* 26: 713-717.
23. Smith, M.L. and Griffith, W.S. 2008. The analysis and comparison of start-up demonstration tests. *European Journal of Operational Research* 186: 1029-1045.
24. Vellaisamy, P. 1996. On the number of successes in dependent trials. *Communications in Statistics-Theory and Methods* 25: 1745-1756.
25. Vellaisamy, P. and Sankar, S. 2005. A unified approach for modeling and designing attribute sampling plans for monitoring dependent production processes. *Methodology and Computing in Applied Probability* 7: 307-323.
26. Vellaisamy, P. and Upadhye, N.S. 2007. On the negative binomial distribution and its generalizations. *Statistics & Probability Letters* 77: 173-180.
27. Yalcin, F. and Eryilmaz, S. 2012. Start-up demonstration test based on total successes and total failures with dependent start-ups. *IEEE Transactions on Reliability* 61: 227-230.
28. Yu, C. and Zelterman, D. 2008. Sums of exchangeable Bernoulli random variables for family and litter frequency data. *Computational Statistics and Data Analysis* 52: 1636-1649.
29. Zaigraev, A. and Kaniovski, S. 2010. Exact bounds on the probability of at least k successes in n exchangeable Bernoulli trials as a function of correlation coefficients. *Statistics & Probability Letters* 80: 1079-1084.

8 Applications

This section is devoted to applications of the discrete stochastic models that have been presented in the preceding sections. First, the use of the discrete stochastic models for evaluating and computing system reliability is discussed with illustrative examples. The methods for computing reliability of coherent systems are presented. Particular class of shock models is defined and evaluated via phase-type and matrix-geometric distributions. Some age replacement policies are presented under discrete setup. The usefulness of discrete stochastic models in statistical quality control applications is discussed for the evaluation and analysis of control charts and sampling plans. Finally, start-up demonstration tests are defined and their characteristics are evaluated.

8.1 SYSTEM RELIABILITY MODELS

The reliability of a system is the probability that the system will perform satisfactorily for at least a given period of time when used under stated conditions (Kuo and Zuo, 2003). For a given system structure, one of the main problems in reliability theory is the computation of system reliability based on component reliabilities. As stated in section 6.2, for a binary coherent system that consists of n components with reliabilities $p_1,...,p_n$, the reliability of the overall system is given by

$$R(p_1,...,p_n) = P\{\phi(X_1,...,X_n) = 1\}, \tag{8.1}$$

where $p_i = P\{X_i = 1\}$ is the reliability of the ith component, $i = 1,...,n$, and ϕ is the structure function of the system.

For the series and parallel structures consisting of n independent components, the reliabilities can be respectively computed from

$$R_{\text{series}}(p_1,...,p_n) = \prod_{i=1}^{n} p_i,$$

and

$$R_{parallel}(p_1,...,p_n) = 1 - \prod_{i=1}^{n}(1-p_i).$$

Numerous methods have been developed to compute the reliability defined by (8.1). Some well-known techniques are combinatorial method, Markov chain imbedding approach and universal generating function method. Manifestly, the choice of the suitable method heavily depends on the structure of the system and the conditions on the system and its components. Combinatorial methods have been effectively used in numerous cases when the system consists of independent and identical components. In this context, the computation of the system reliability is directly related to the number of binary sequences of length n that ensures the operation of the

DOI: 10.1201/9781003160632-8

system. See, for example the system considered in Example 4.2 and Table 4.1 which lists all possible sequences that ensure the functioning of the system.

In the following, the reliabilities of all binary coherent systems of order $n = 3$ are computed by considering all possible components' states that ensure the functioning of the system. The components are assumed to be independent and the reliability of the ith component is given by $p_i, i = 1,2,3$. There are five binary coherent systems of order $n = 3$ which are listed below.

1. $\phi(x_1,x_2,x_3) = \min(x_1,x_2,x_3)$

$$\begin{aligned} R(p_1,p_2,p_3) &= P\{X_1 = 1, X_2 = 1, X_3 = 1\} \\ &= p_1p_2p_3. \end{aligned}$$

2. $\phi(x_1,x_2,x_3) = \min(x_2,\max(x_1,x_3))$

$$\begin{aligned} R(p_1,p_2,p_3) &= P\{X_1 = 1, X_2 = 1, X_3 = 0\} \\ &\quad + P\{X_1 = 0, X_2 = 1, X_3 = 1\} + P\{X_1 = 1, X_2 = 1, X_3 = 1\} \\ &= p_1p_2(1-p_3) + (1-p_1)p_2p_3 + p_1p_2p_3 \\ &= p_1p_2 + p_2p_3 - p_1p_2p_3. \end{aligned}$$

3. $\phi(x_1,x_2,x_3) = \min(\max(x_1,x_2),\max(x_1,x_3),\max(x_2,x_3))$

$$\begin{aligned} R(p_1,p_2,p_3) &= P\{X_1 = 1, X_2 = 1, X_3 = 0\} + P\{X_1 = 1, X_2 = 0, X_3 = 1\} \\ &\quad + P\{X_1 = 0, X_2 = 1, X_3 = 1\} + P\{X_1 = 1, X_2 = 1, X_3 = 1\} \\ &= p_1p_2(1-p_3) + p_1(1-p_2)p_3 + (1-p_1)p_2p_3 + p_1p_2p_3 \\ &= p_1p_2 + p_1p_3 + p_2p_3 - 2p_1p_2p_3. \end{aligned}$$

4. $\phi(x_1,x_2,x_3) = \max(x_2,\min(x_1,x_3))$

$$\begin{aligned} R(p_1,p_2,p_3) &= P\{X_1 = 1, X_2 = 0, X_3 = 1\} + P\{X_1 = 1, X_2 = 1, X_3 = 0\} \\ &\quad + P\{X_1 = 0, X_2 = 1, X_3 = 0\} + P\{X_1 = 0, X_2 = 1, X_3 = 1\} \\ &\quad + P\{X_1 = 1, X_2 = 1, X_3 = 1\} \\ &= p_1(1-p_2)p_3 + p_1p_2(1-p_3) + (1-p_1)p_2(1-p_3) \\ &\quad + (1-p_1)p_2p_3 + p_1p_2p_3 \\ &= p_2 + p_1p_3 - p_1p_2p_3. \end{aligned}$$

5. $\phi(x_1,x_2,x_3) = \max(x_1,x_2,x_3)$

$$\begin{aligned} R(p_1,p_2,p_3) &= 1 - P\{X_1 = 0, X_2 = 0, X_3 = 0\} \\ &= 1 - (1-p_1)(1-p_2)(1-p_3). \end{aligned}$$

Consider the reliability of the system defined by the structure function

$$\phi(x_1,x_2,x_3) = \min(\max(x_1,x_2),\max(x_1,x_3),\max(x_2,x_3))$$

when the components are exchangeable dependent. Under exchangeability, we have

$$\begin{aligned}P\{X_1=1,X_2=1,X_3=0\} &= P\{X_1=1,X_2=0,X_3=1\}\\ &= P\{X_1=0,X_2=1,X_3=1\}.\end{aligned}$$

Thus, the system reliability can be represented as

$$R=3P\{X_1=1,X_2=1,X_3=0\}+P\{X_1=1,X_2=1,X_3=1\}.$$

Similarly, for the system with structure function

$$\phi(x_1,x_2,x_3)=\max(x_2,\min(x_1,x_3)),$$

if the components are exchangeable, then its reliability is

$$\begin{aligned}R=&P\{X_1=0,X_2=1,X_3=0\}+3P\{X_1=1,X_2=1,X_3=0\}\\ &+P\{X_1=1,X_2=1,X_3=1\}.\end{aligned}$$

Example 8.1. Consider a series system with two components denoted by "1" and "2". Two identical redundant components denoted by "3" and "4" are used to increase the reliability of the system. The redundancy can be applied at component or system level. The structure function of the system has the following form when system level redundancy is applied.

$$\phi_1(x_1,x_2,x_3,x_4)=\max(\min(x_1,x_2),\min(x_3,x_4)).$$

On the other hand, if component level redundancy is applied, then the structure function becomes

$$\phi_2(x_1,x_2,x_3,x_4)=\min(\max(x_1,x_3),\max(x_2,x_4)).$$

Assume that the components are independent and have the same reliability value p. Then, the reliability of the system under the system level redundancy is

$$\begin{aligned}R_1(p)&=P\{\phi_1(X_1,X_2,X_3,X_4)=1\}\\ &=1-(1-p^2)(1-p^2)\\ &=2p^2-p^4.\end{aligned}$$

Under the component level redundancy,

$$\begin{aligned}R_2(p)&=P\{\phi_2(X_1,X_2,X_3,X_4)=1\}\\ &=\left[1-(1-p)^2\right]\left[1-(1-p)^2\right]\\ &=4p^2-4p^3+p^4.\end{aligned}$$

It can be easily seen that

$$R_2(p)-R_1(p)=2p^2(1-p)^2>0.$$

Thus, for the corresponding series system, adding redundancy at the component level is better in terms of reliability. ■

The computation of (8.1) based on $p_1,...,p_n$ for a given coherent system is a discrete probabilistic problem and the stochastic models that have been presented throughout this book can be effectively used for this purpose.

Example 8.2. The k-out-of-n system model is one of the basic models in system reliability. Such a system consists of n components and functions if at least k components work. This reliability model has been extensively studied in the literature under various assumptions on the system and its components. Suppose that the system has n independent and nonidentical components. That is, the components within the system have different reliability values denoted by $p_1,...,p_n$. The structure function of this system can be defined as

$$\phi(x_1,...,x_n) = \begin{cases} 1, & \text{if } \sum_{i=1}^n x_i \geq k \\ 0, & \text{if } \sum_{i=1}^n x_i < k, \end{cases}$$

where $x_i = 1$ if the ith component is functioning and $x_i = 0$ if the ith component has failed, $i = 1,...,n$.

If $S_n = \sum_{i=1}^n X_i$ denotes the total number of working components, then the reliability of the k-out-of-n system can be represented as

$$R_k(p_1,...,p_n) = P\{S_n \geq k\}. \tag{8.2}$$

Because the system has independent and nonidentical components, as pointed out in section 2, the random variable S_n has Poisson's Binomial distribution. Thus, a recursive equation can be obtained to compute (8.2) for given n and k. Clearly, $P\{S_n \geq k\} = 0$ if $n < k$ and $P\{S_n \geq k\} = 1$ if $k = 0$. On the other hand, if $n \geq k$ and $k \geq 1$, then

$$P\{S_n \geq k\} = P\{S_{n-1} \geq k-1\} p_n + P\{S_{n-1} \geq k\}(1-p_n).$$

Thus, we have

$$R_k(p_1,...,p_n) = R_{k-1}(p_1,...,p_{n-1})p_n + R_k(p_1,...,p_{n-1})(1-p_n), \tag{8.3}$$

for $n \geq k$ and $k \geq 1$ with $R_k(p_1,...,p_n) = 0$ for all $n < k$ and $R_0(p_1,...,p_n) = 1$ for all n. If the system has independent and identical components, i.e. $p_i = p$, then the system reliability can be computed via Binomial distribution. Indeed,

$$P\{S_n \geq k\} = \sum_{s=k}^{n} \binom{n}{s} p^s (1-p)^{n-s}. \tag{8.4}$$

The recursive equation (8.3) can also be used to compute the reliability of the system with identical components. However, the equation (8.3) is computationally more complex than (8.4). ■

If the components of the system described in Example 8.2 are exchangeable dependent, then using equation (7.2), the reliability of the system can be calculated from

$$P\{S_n \geq k\} = \sum_{s=k}^{n} P\{S_n = s\}$$
$$= \sum_{s=k}^{n} \binom{n}{s} \sum_{i=0}^{n-s} (-1)^i \binom{n-s}{i} \lambda_{s+i},$$

where $\lambda_0 = 1$ and $\lambda_m = P\{X_1 = 1, ..., X_m = 1\}, m = 1, 2, ...$ Clearly, λ_m denotes the joint working probability of $m \geq 1$ components. For a given system structure, a component can be more important than others in terms of its contribution to overall system reliability. Measuring the importance of a component within the system is useful in reliability based design. For a system consisting of n components, the Birnbaum importance of the ith component is defined by

$$I_i = P\{\text{System functions} \mid X_i = 1\} - P\{\text{System functions} \mid X_i = 0\},$$

for $i = 1, 2, ..., n$ (Birnbaum 1969). The Birnbaum importance measures the difference between conditional reliabilities of the system when the ith component is functioning and when it has failed. There are other types of component importance measures. See, e.g. (Kuo and Zhu 2012) for a comprehensive review of reliability importance measures. Reliability based design can be established depending on the values of $I_1, I_2, ..., I_n$. For example, under certain conditions, the redundancy can be applied to the component which has the highest importance value.

Example 8.3. Consider the system with structure function

$$\phi(x_1, x_2, x_3) = \min(x_1, \max(x_2, x_3)).$$

Assume that the components are independent and $p_i = P\{X_i = 1\}, i = 1, 2, 3$. The Birnbaum importance of component "1" is

$$I_1 = P\{\text{System functions} \mid X_1 = 1\} - P\{\text{System functions} \mid X_1 = 0\}.$$

Manifestly,

$$\begin{aligned}
&P\{\text{System functions}, X_1 = 1\} \\
&= P\{X_1 = 1, X_2 = 1, X_3 = 1\} \\
&+ P\{X_1 = 1, X_2 = 1, X_3 = 0\} \\
&+ P\{X_1 = 1, X_2 = 0, X_3 = 1\} \\
&= p_1 p_2 p_3 + p_1 p_2 (1 - p_3) + p_1 (1 - p_2) p_3.
\end{aligned}$$

Thus, we have

$$\begin{aligned}
&P\{\text{System functions} \mid X_1 = 1\} \\
&= \frac{p_1 p_2 p_3 + p_1 p_2 (1 - p_3) + p_1 (1 - p_2) p_3}{p_1} \\
&= p_2 + p_3 - p_2 p_3,
\end{aligned}$$

and

$$P\{\text{System functions} \mid X_1 = 0\} = 0.$$

Therefore, the Birnbaum importance of component "1" is found to be

$$I_1 = p_2 + p_3 - p_2 p_3.$$

Similarly,

$$I_2 = p_1(1 - p_3),$$

and

$$I_3 = p_1(1 - p_2).$$

Let $p_1 = p_2 = p_3 = p$. Then,

$$I_1 = 2p - p^2,$$
$$I_2 = p - p^2,$$
$$I_3 = p - p^2.$$

Thus, if the components are identical, then

$$I_1 > I_2 = I_3.$$

Example 8.4. Consider an oil transportation system for transporting oil from the point A to the point B. There are n equally spaced pump stations between the points A and B. Depending on the transportation capacity of the pumps, the following systems are considered. System 1 uses $n = 30$ pumps of type I and each pump is able to transport oil to the next pump. Thus, if any one of the pumps fails, then the system cannot transport the oil. System 2 uses $n = 30$ pumps of type II and each pump is able to transport oil to a distance including three other pumps. Therefore, the system fails if and only if at least three consecutive pumps fail. Manifestly, System 1 has a series structure with $n = 30$ components. On the other hand, System 2 has a linear consecutive-3-out-of-30:F structure. Note that a linear consecutive-k-out-of-n:F system is a system that consists of n linearly ordered components and fails if at least k consecutive components fail (see, e.g. (Eryilmaz 2010a)). Assume that the failure probability of a pump of type 1 is 0.003 while the failure probability of each pump of type 2 is 0.03. Then, the reliability of the System 1 is $(1 - 0.003)^{30} = 0.9138$. On the other hand, the reliability of the System 2 is 0.9993 (see Example 2.11). That is, System 2 is more reliable than System 1 although the components within System 1 has a larger reliability (smaller failure probability). ■

Example 8.4 clearly exhibits the importance of the optimal design in terms of reliability. If only the reliability of the system is considered as a main criteria for choosing the better system design, depending on the calculations in Example 8.4, System 2 should be setup. That is, pumps of type II should be used in the construction of the oil transportation system. However, the pumps of type II might be more expensive than pumps of type I. Thus, the optimal design problem should consider both reliability and cost values. Indeed, for a manufactured system, there are two main questions that need to be answered. First, "How much does it cost?", second

"How well does it perform?". Clearly, the latter one is related to the reliability of the system.

Example 8.5. Consider a series-parallel system that consists of K parallel systems such that the ith parallel system has n_i+1 independent components which have common unreliability (failure probability) value q_i, $i=1,2,...,K$. The problem is to find the optimal number of redundancies, i.e. the optimum value of $(n_1,n_2,...,n_K)$ that maximizes the reliability of the entire series-parallel system under a cost constraint. That is, we have the following optimization problem:

$$\max R(\mathbf{n}) = \prod_{i=1}^{K}(1-q_i^{n_i+1})$$

subject to

$$\sum_{i=1}^{K} c_i n_i \leq C,$$

where c_i is the cost of components within the ith parallel system, and C is a given upper limit for the total cost when n_i components are used in the ith parallel system. An approach for the solution of this problem is based on the determination of the increment in the logarithm of the system reliability resulting from adding a single unit of type i (Barlow and Proschan 1996). Because $\log x$ is a monotone increasing function of x, maximizing $\log R(\mathbf{n})$ is equivalent to maximizing $R(\mathbf{n})$. Starting with the cheapest cost allocation $(0,0,...,0)$ more expensive allocations are obtained successively. If the present allocation is $\mathbf{n}$, the index i_0 is determined for which

$$\frac{1}{c_i}\left[\log(1-q_i^{n_i+2})-\log(1-q_i^{n_i+1})\right]$$

is maximum over $i=1,2,...,K$. The next allocation becomes $n_1,...,n_{i_0-1},n_{i_0}+1,n_{i_0+1},...,n_K$. The detail of the algorithm is presented in (Barlow and Proschan 1996).

Example 8.6. The reliability of the linear consecutive-k-out-of-n:F system that consists of independent and identical components can also be computed via phase-type distributions. For the survival of the system, no k or more consecutive failures must occur in a binary sequence of size n. Thus, if T_k denotes the number of trials until k consecutive failures in a sequence of Bernoulli trials, then the probability $P\{T_k>n\}$ gives the reliability of the linear consecutive-k-out-of-n:F system. From Example 4.4, we have $T_k \sim PH_k(\pi,\mathbf{Q})$, where $\pi=(1,0,...,0)$, and the $k\times k$ matrix $\mathbf{Q}$ is given by

$$\mathbf{Q}=\begin{bmatrix} p & 1-p & 0 & \dots & 0 \\ p & 0 & 1-p & \dots & 0 \\ \vdots & \vdots & \vdots & & \vdots \\ p & 0 & 0 & \dots & 0 \end{bmatrix},$$

where p is the common reliability of components. Note that the p in Example 4.4 is replaced by $1-p$ since in the present setup T_k denotes the number of trials until k consecutive failures. Thus, the reliability of the linear consecutive-k-out-of-n:F system is

$$R(p) = \pi \mathbf{Q}^n \mathbf{e}'.$$

This method is closely related to the Markov chain approach for the study of reliability structures. See, e.g. (Koutras 1996). For $2k \geq n$, the reliability of the linear consecutive-k-out-of-n:F system can be computed by the following simple equation (Tong 1985):

$$R(p) = 1 - (n-k+1)(1-p)^k + (n-k)(1-p)^{k+1}.$$

Table 8.1 contains reliability values of the linear consecutive-k-out-of-n:F system for selected values of k, n, and p. The reliability of the system is increasing in k, p, and decreasing in n.

Table 8.1
Reliability of the Linear Consecutive-*k*-out-of-*n*:F system

n	*k*	$p = 0.95$	$p = 0.97$
10	2	0.9786	0.9921
	3	0.9999	0.9998
20	2	0.9555	0.9835
	3	0.9979	0.9995
30	2	0.9330	0.9749
	3	0.9967	0.9993
40	2	0.9109	0.9664
	3	0.9955	0.9990
50	2	0.8894	0.9580
	3	0.9943	0.9987

If $R(k,n)$ denotes the reliability of the linear consecutive-k-out-of-n:F system, then the Birnbaum importance of the ith component in the linear consecutive-k-out-of-n: F system can be calculated from

$$I_i = \frac{R(k,i-1)R'(k,n-i) - R(k,n)}{1-p_i},$$

where p_i is the reliability of the ith component, and $R'(k,m)$ is the reliability of the consecutive-k-out-of-m:F subsystem consisting of components $n-m+1, n-m+2, \ldots, n-1, n$ (Papastavridis 1987). The Birnbaum importance ordering of the components is of special importance in optimal design of the linear consecutive-k-out-of-n:F system. We have n components with respective reliabilities $p_1, \ldots, p_n$ and wish to find the optimal assignment of the components to the n nodes to maximize the system reliability. Consider the linear consecutive-2-out-of-3:F system. Assume that we have the components with reliabilities 0.96, 0.97, and 0.94. The reliability of the system is

$$R(2,3) = p_2 + p_1 p_3 - p_1 p_2 p_3.$$

Below, we compute $R(2,3)$ for each possible arrangement. Note that if we have the arrangement 132, then $p_1 = 0.96, p_2 = 0.94$, and $p_3 = 0.97$.

$$\begin{aligned} 123 &\Longrightarrow R(2,3) = 0.9971, \\ 132 &\Longrightarrow R(2,3) = 0.9959, \\ 213 &\Longrightarrow R(2,3) = 0.9965, \\ 231 &\Longrightarrow R(2,3) = 0.9959, \\ 312 &\Longrightarrow R(2,3) = 0.9965, \\ 321 &\Longrightarrow R(2,3) = 0.9971. \end{aligned}$$

Among these six possible arrangements, the arrangements 123 and 321 are optimal. Indeed, $I_2 > I_3$ for $p_2 > p_3$ and $I_1 < I_2$ for $p_1 < p_2$.

(Chadjiconstantinidis and Koutras 1999) presented expressions for the Birnbaum importance of the components for the wide class of Markov chain embeddable systems.

Example 8.7. For a system consisting of a specified number of components, the number X of components that are failed at the time of system failure is a useful quantity, and gives an idea of how many spares should be available to replace all failed components when the system fails (Papastavridis 1989). If the components have continuous failure time distributions, then for series and parallel systems we respectively have $P\{X = 1\} = 1$ and $P\{X = n\} = 1$. For the k-out-of-n system, $P\{X = n-k+1\} = 1$. Consider the system with the lifetime $T = \min(T_1, \max(T_2, T_3))$, where T_1, T_2, and T_3 denote the failure times (lifetimes) of the components. In Table 8.2, we present the value of X for each possible ordering between components lifetimes.

Table 8.2
The Values of X for all Possible Orderings between Components' Lifetimes

Ordering	X
$T_1 < T_2 < T_3$	1
$T_1 < T_3 < T_2$	1
$T_2 < T_1 < T_3$	2
$T_2 < T_3 < T_1$	2
$T_3 < T_1 < T_2$	2
$T_3 < T_2 < T_1$	2

Thus, from Table 8.2, if the components are independent and have identical continuous failure time distribution, then $P\{X = 1\} = \frac{2}{6}$ and $P\{X = 2\} = \frac{4}{6}$. For a general coherent system with lifetime T and signature vector $\mathbf{s} = (s_1, ..., s_n)$,

$$P\{X = i\} = P\{T = T_{i:n}\} = s_i,$$

$i = 1, ..., n$, where $T_{i:n}$ is the ith smallest among $T_1, ..., T_n$ (Eryilmaz 2012). Therefore, the mean number of failed components that are failed at the time of system failure is in fact the expected value of the discrete distribution $\mathbf{s} = (s_1, ..., s_n)$, i.e.

$$E(X) = \sum_{i=1}^{n} i \cdot s_i. \quad \blacksquare$$

Below, we present the signature of all coherent systems of order $n = 3$ and compute the value of $E(X)$:

1. $\phi(x_1, x_2, x_3) = \min(x_1, x_2, x_3)$

$$\mathbf{s} = (1,0,0) \ , E(X) = 1$$

2. $\phi(x_1, x_2, x_3) = \min(x_2, \max(x_1, x_3))$

$$\mathbf{s} = \left(\frac{1}{3}, \frac{2}{3}, 0\right) \ , E(X) = 1.6667$$

3. $\phi(x_1, x_2, x_3) = \min(\max(x_1, x_2), \max(x_1, x_3), \max(x_2, x_3))$

$$\mathbf{s} = (0,1,0) \ , E(X) = 2$$

4. $\phi(x_1, x_2, x_3) = \max(x_2, \min(x_1, x_3))$

$$\mathbf{s} = \left(0, \frac{2}{3}, \frac{1}{3}\right) \ , E(X) = 2.3333$$

5. $\phi(x_1, x_2, x_3) = \max(x_1, x_2, x_3)$

$$\mathbf{s} = (0,0,1) \ , E(X) = 3.$$

The signature of all coherent systems with five or less components were computed and presented by (Navarro and Rubio 2010).

Example 8.8. A linear consecutive k-within-m-out-of-n:F system is a system that consists of n linearly ordered components such that the system fails iff there are m consecutive components which include among them at least k failed components (Malinowski and Preuss 1985). In practice, this system model has applications in quality control, inspection procedures and radar detection. For $k = 2$, the signature of the linear consecutive 2-within-m-out-of-n:F system is

$$s_i = \binom{n}{i}^{-1} \left[\frac{n-i+1}{i} \binom{n-(i-2)(m-1)}{i-1} - \binom{n-(i-1)(m-1)}{i} \right],$$

for $i = 1, ..., n$ (Eryilmaz 2010b). In Table 8.3, the mean number of failed components that are failed at the time when the consecutive-2-within-m-out-of-n:F system fails is computed for selected values of m and n.

In Examples 8.7 and 8.8, the distribution of the discrete random variable X has been computed for the case when the systems have continuous distribution for the

Table 8.3
The Mean Number of Failed Component at the Time when the Consecutive-2-within-*m*-out-of-*n*:F System Fails

n	*m*	*E*(*X*)
10	3	2.7937
	4	2.5000
	5	2.3333
20	3	3.5597
	4	3.1068
	5	2.8390
	6	2.6589

components' failure times. As stated above, if the component failure time distribution is continuous, then we have $P\{X=n-k+1\}=1$ for the k-out-of-n system. However, this may not be the case when the components have discrete failure time distribution. This is due to positive probability of ties between times of component failures. (Davies and Dembinska 2019) obtained expression for the distribution of the number of failed components in a k-out-of-n system upon system failure when the lifetimes are discretely distributed. Let the components have discrete lifetime distribution with

$$F(t)=P\{T_i\le t\}\ ,\ F(t^-)=P\{T_i<t\},\ p(t)=P\{T_i=t\},$$

for $i=1,2,\ldots,n$. Then, (Davies and Dembinska 2019) obtained the following equation:

$$\begin{aligned}
&P\{X=n-k+m+1,T_{n-k+1:n}=t\}\\
&=\frac{n!}{(k-m-1)!}\left(\bar{F}(t)\right)^{k-m-1}\\
&\times\sum_{s=0}^{n-k}\frac{1}{s!(n-k+m+1-s)!}\left(F(t^-)\right)^s\left(p(t)\right)^{n-k+m+1-s},
\end{aligned}$$

for $m=0,1,\ldots,k-1$. Therefore, the distribution of the number of failed components at the time when the k-out-of-n system fails can be computed from

$$P\{X=n-k+m+1\}=\sum_{t=1}^{\infty}P\{X=n-k+m+1,T_{n-k+1:n}=t\}.$$

Assume that the system has a parallel structure, i.e. $k=1$. Then,

$$\begin{aligned}
&P\{X=n,T_{n:n}=t\}\\
&=\sum_{s=0}^{n-1}\binom{n}{s}\left(F(t^-)\right)^s\left(p(t)\right)^{n-s},\\
&=\left(F(t^-)+p(t)\right)^n-\left(F(t^-)\right)^n\\
&=\left(F(t)\right)^n-\left(F(t^-)\right)^n\\
&=P\{T_{n:n}=t\}.
\end{aligned}$$

Thus, for the parallel system

$$\begin{aligned} P\{X=n\} &= \sum_{t=1}^{\infty} P\{X=n, T_{n:n}=t\} \\ &= \sum_{t=1}^{\infty} P\{T_{n:n}=t\} \\ &= 1. \end{aligned}$$

That is, even in the discrete case, the number of failed components at the time when the parallel system fails is n as in the case of continuous failure time distributions. This is an expected result. However, this may not be true for any k-out-of-n system. Let $k=n$, i.e. the system has a series structure. Then,

$$\begin{aligned} &P\{X=m+1, T_{1:n}=t\} \\ &= \binom{n}{m+1} (p(t))^{m+1} (\bar{F}(t))^{n-m-1}, \end{aligned}$$

for $m=0,1,...,n-1$, and

$$\begin{aligned} &P\{X=m+1\} \\ &= \binom{n}{m+1} \sum_{t=1}^{\infty} (p(t))^{m+1} (\bar{F}(t))^{n-m-1}. \end{aligned}$$

Example 8.9. For a series system that consists of n independent and identical components, assume that the lifetimes of the components have geometric distribution with

$$p(t) = P\{T_i = t\} = p(1-p)^{t-1} \text{ and } F(t) = P\{T_i \le t\} = 1-(1-p)^t,$$

for $t=1,2,....$ Then,

$$\begin{aligned} &P\{X=m+1\} \\ &= \binom{n}{m+1} \sum_{t=1}^{\infty} (p(1-p)^{t-1})^{m+1} ((1-p)^t)^{n-m-1} \\ &= \binom{n}{m+1} \frac{p^{m+1}(1-p)^{n-m-1}}{1-(1-p)^n}, \end{aligned}$$

for $m=0,1,...,n-1$. The mean number of failed components at the time when the series system fails can be computed as

$$E(X) = \frac{np}{1-(1-p)^n}.$$

For an illustration, let $n = 10$ and $p = 0.4$. Then,

$$P\{X=1\}=0.0405,\ P\{X=2\}=0.1217,$$
$$P\{X=3\}=0.2163, P\{X=4\}=0.2523,$$
$$P\{X=5\}=0.2019,\ P\{X=6\}=0.1122,$$
$$P\{X=7\}=0.0427,\ P\{X=8\}=0.0107,$$
$$P\{X=9\}=0.0016,\ P\{X=10\}=0.0001,$$

and

$$E(X)=\frac{10p}{1-(1-p)^{10}}=4.0243.$$

Therefore, on average, the number of failed components upon the system failure is around 4.

Example 8.10. In section 6, for a binary coherent system with n independent and identical components with common reliability p, the reliability of the system has been expressed as

$$R(p)=\sum_{i=1}^{n}s_i\sum_{j=n-i+1}^{n}\binom{n}{j}p^j(1-p)^{n-j}, \tag{8.5}$$

where $\mathbf{s}=(s_1,...,s_n)$ is the signature of the system. The reliabilities of two systems having different structures can be compared using signature vectors. For two discrete distributions $\mathbf{s}=(s_1,...,s_n)$ and $\mathbf{s}^*=(s_1^*,...,s_n^*)$,

$$\mathbf{s}\leq_{st}\mathbf{s}^* \text{ if } \sum_{j=i}^{n}s_j\leq\sum_{j=i}^{n}s_j^* \text{ for all } i=1,...,n.$$

Let $R_1(p)$ $(R_2(p))$ denote the reliability of the system with signature $\mathbf{s}$ $(\mathbf{s}^*)$. Both systems use same types of components. Therefore, the reliability values of the components in each system is p. If $\mathbf{s}\leq_{st}\mathbf{s}^*$, then $R_1(p)\leq R_2(p)$ (Samaniego 2007). ■

An alternative expression for (8.5) is given by

$$R(p)=\sum_{i=1}^{n}\alpha_i p^i,$$

where the vector of coefficients $(\alpha_1,...,\alpha_n)$ satisfying $\sum_{i=1}^{n}\alpha_i=1$ is called minimal signature (Navarro et al. 2007). The following relation holds true between the minimal signature and the signature of a given coherent structure with n independent and identical components.

$$\alpha_i=\sum_{l=n-i}^{n-k_{\min}}(-1)^{i+l-n}\binom{l}{n-i}\binom{n}{l}\sum_{j=l+1}^{n}s_j, \tag{8.6}$$

for $i\geq k_{\min}$, and $\alpha_i=0$ for $i<k_{\min}$, where $k_{\min}$ is the minimum number of working components in a coherent structure such that the system can still operate successfully (Eryilmaz 2010b).

Example 8.11. Assume that three independently working high speed cameras are used at slightly different angles focusing on a specific area. The cameras are labeled as 1, 2, and 3. At least two adjacent cameras must work properly to obtain the required data/image corresponding to the reactions in the area (Chang et al. 2000). In this case, we have a system that consists of $n = 3$ components (cameras), and the system functions if and only if at least $k = 2$ consecutive components function. Such a system is a linear consecutive-k-out-of-n:G system. For $n = 3$ and $k = 2$, the structure function of the system is

$$\phi(x_1, x_2, x_3) = \max(\min(x_1, x_2), \min(x_2, x_3)).$$

If the cameras have the common reliability value p, then the system reliability is calculated as

$$\begin{aligned} R(p) &= P\{X_1 = 1, X_2 = 1, X_3 = 0\} + P\{X_1 = 0, X_2 = 1, X_3 = 1\} \\ &\quad + P\{X_1 = 1, X_2 = 1, X_3 = 1\} \\ &= 2p^2(1-p) + p^3 \\ &= 2p^2 - p^3. \end{aligned}$$

Thus, the minimal signature of the system is found to be

$$(\alpha_1, \alpha_2, \alpha_3) = (0, 2, -1). \quad \blacksquare$$

Below, we present the minimal signature of all coherent systems of order $n = 3$.

1. $\phi(x_1, x_2, x_3) = \min(x_1, x_2, x_3)$

$$(\alpha_1, \alpha_2, \alpha_3) = (0, 0, 1)$$

2. $\phi(x_1, x_2, x_3) = \min(x_2, \max(x_1, x_3))$

$$(\alpha_1, \alpha_2, \alpha_3) = (0, 2, -1)$$

3. $\phi(x_1, x_2, x_3) = \min(\max(x_1, x_2), \max(x_1, x_3), \max(x_2, x_3))$

$$(\alpha_1, \alpha_2, \alpha_3) = (0, 3, -2)$$

4. $\phi(x_1, x_2, x_3) = \max(x_2, \min(x_1, x_3))$

$$(\alpha_1, \alpha_2, \alpha_3) = (1, 1, -1)$$

5. $\phi(x_1, x_2, x_3) = \max(x_1, x_2, x_3)$

$$(\alpha_1, \alpha_2, \alpha_3) = (3, -3, 1).$$

(Gertsbakh et al. 2011), (Da et al. 2012), and (Eryilmaz 2014) obtained formulas for computing signatures of series and parallel systems consisting of disjoint modules. Consider a coherent system consisting of two disjoint modules and a total of

$n+m$ components such that the first module with n components has structure function ϕ_1 and the second module with m components has structure function ϕ_2. Let $\mathbf{s}^{(1)} = (s_1^{(1)},...,s_n^{(1)})$ and $\mathbf{s}^{(2)} = (s_1^{(2)},...,s_m^{(2)})$ be the signature vectors of the modules which respectively have structure functions ϕ_1 and ϕ_2. If the system that consists of $n+m$ components has a series structure, i.e. the two modules are serially connected, then its structure function can be represented as

$$\phi_S(x_1,...,x_{n+m}) = \min(\phi_1(x_1,...,x_n), \phi_2(x_{n+1},...,x_{n+m})).$$

If the modules have a parallel connection, then

$$\phi_P(x_1,...,x_{n+m}) = \max(\phi_1(x_1,...,x_n), \phi_2(x_{n+1},...,x_{n+m})).$$

The problem is to compute the signatures of the systems having structure functions $\phi_S(x_1,...,x_{n+m})$ and $\phi_P(x_1,...,x_{n+m})$ based on the signatures $\mathbf{s}^{(1)}$ and $\mathbf{s}^{(2)}$ of the two disjoint modules. Let $\mathbf{s}_S = (s_{S,1},...,s_{S,n})$ and $\mathbf{s}_P = (s_{P,1},...,s_{P,n})$ be the signature vectors of the systems which respectively have structure functions ϕ_S and ϕ_P. Denote by $k^i_{\min}$ the minimum number of working components for the functioning of the ith module, $i = 1,2$.. If the overall system has a series structure, then (Eryilmaz 2014) has shown that

$$s_{S,i} = \frac{h_{n+m-i+1}(n+m)}{\binom{n+m}{i-1}} - \frac{h_{n+m-i}(n+m)}{\binom{n+m}{i}},$$

where

$$h_i(n+m) = \sum_{j=\max(k^1_{\min},i-m)}^{\min(i-k^2_{\min},n)} \binom{n}{j} \left(\sum_{a=n-j+1}^{n} s_a^{(1)}\right) \binom{m}{i-j} \left(\sum_{b=m-(i-j)+1}^{m} s_b^{(2)}\right).$$

If the overall system has a parallel structure, then (Eryilmaz 2014) has obtained the following equation:

$$s_{P,i} = \frac{c_{n+m-i+1}(n+m)}{\binom{n+m}{i-1}} - \frac{c_{n+m-i}(n+m)}{\binom{n+m}{i}},$$

where

$$\begin{aligned} c_i(n+m) = & \sum_{j=\max(k^1_{\min},i-m)}^{\min(i,n)} \binom{m}{i-j} \binom{n}{j} \left(\sum_{a=n-j+1}^{n} s_a^{(1)}\right) \\ & + \sum_{l=\max(k^2_{\min},i-n)}^{\min(i,m)} \binom{n}{i-l} \binom{m}{l} \left(\sum_{b=m-l+1}^{m} s_b^{(2)}\right) - h_i(n+m). \end{aligned}$$

Example 8.12. Consider the system with the structure function

$$\phi(x_1,x_2,x_3,x_4) = \min(\max(x_1,x_2), \max(x_3,x_4)).$$

Define

$$\phi_1(x_1,x_2) = \max(x_1,x_2),$$

and

$$\phi_2(x_3,x_4) = \max(x_3,x_4).$$

Then,

$$\mathbf{s}^{(1)} = (0,1)\ ,\ \mathbf{s}^{(2)} = (0,1).$$

Using the signature vectors $\mathbf{s}^{(1)}$ and $\mathbf{s}^{(2)}$, the signature of the system with structure function $\phi(x_1,x_2,x_3,x_4) = \min(\max(x_1,x_2),\max(x_3,x_4))$ is found to be

$$\mathbf{s}_S = (0,\frac{1}{3},\frac{2}{3},0). \quad \blacksquare$$

In a traditional setup of reliability problems, all components within the system are assumed to have identical reliability values. This may not be true for real-life systems. The concept of survival signature has been found to be useful to study coherent systems that consist of multiple types of components. In the following example, the reliability function of a coherent system consisting of K different types of components is represented via survival signature.

Example 8.13. Consider a coherent system with n components. Let n_i denote the number of components of type i, $i = 1,2,...,K$, where $n = \sum_{i=1}^{K} n_i$. Assume that the components are independent and p_i denotes common reliability of components of type i. In this case, by conditioning on the number of working components of each type, the reliability function of the system can be represented as

$$R(p_1,...,p_K) = \sum_{l_1=0}^{n_1} \cdots \sum_{l_K=0}^{n_K} \Phi(l_1,...,l_k) \prod_{i=1}^{K} \binom{n_i}{l_i} p_i^{l_i}(1-p_i)^{n_i-l_i},$$

where $\Phi(l_1,...,l_k)$ denotes the survival signature which is defined by

$$\Phi(l_1,...,l_k) = \frac{r_{n_1,....,n_K}(l_1,...,l_K)}{\binom{n_1}{l_1}\cdots\binom{n_K}{l_K}},$$

where $r_{n_1,....,n_K}(l_1,...,l_K)$ denotes the number of path sets of the system including exactly l_1 components of type 1,..., exactly l_K components of type K (Coolen and Coolen-Maturi 2012), (Coolen and Coolen-Maturi 2015). $\blacksquare$

Example 8.14. Consider a consecutive k-out-of-n:F system that consists of n linearly ordered components such that the first n_1 components are of type I and the remaining $n_2 = n - n_1$ components are of type II. The reliability of each component of type i is $p_i, i = 1,2$. (Eryilmaz 2016) obtained the survival signature for this system. In particular, for $l_1, l_2 > 0$

$$r_{n_1,n_2}(l_1,l_2) = C_k(l_1+1,n_1-l_1)C_k(l_2,n_2-l_2)$$
$$+ \sum_{x=0}^{k-2} \sum_{y=1}^{k-x-1} C_k(l_1,n_1-l_1-x)C_k(l_2,n_2-l_2-y),$$

$$r_{n_1,n_2}(l_1,0) = \sum_{x=0}^{k-n_2-1} C_k(l_1, n_1 - l_1 - x), \text{ for } k > n_2,$$

and

$$r_{n_1,n_2}(0,l_2) = \sum_{y=0}^{k-n_1-1} C_k(l_2, n_1 - l_2 - y), \text{ for } k > n_1,$$

where $r_{n_1,n_2}(0,0) = 0$, and

$$C_k(a,b) = \sum_{m=0}^{\min(a,[\frac{b}{k}])} (-1)^m \binom{a}{m}\binom{a+b-mk-1}{a-1},$$

for $a > 0, b \geq 0$, and $C_k(a,b) = 1$ if $a \geq 0$ and $b = 0$. For an illustration, let $k = 2, n = 10$, and $n_1 = 3$. Then, $\Phi(l_1,l_2)$ is computed for some selected values of l_1 and l_2 :

$$\Phi(1,5) = \frac{10}{63},$$
$$\Phi(2,5) = \frac{40}{63},$$
$$\Phi(3,5) = \frac{5}{7},$$
$$\Phi(1,6) = \frac{2}{7},$$
$$\Phi(2,6) = \frac{20}{21},$$
$$\Phi(3,6) = 1.$$

If for example, $p_1 = 0.95$ and $p_2 = 0.97$, then the reliability of the consecutive 2-out-of-10:F system when the first $n_1 = 3$ components are of type I and the remaining $n_2 = 7$ components are of type II is

$$\begin{aligned} R(p_1,p_2) &= \sum_{l_1=0}^{n_1}\sum_{l_2=0}^{n_2} \Phi(l_1,l_2)\prod_{i=1}^{2}\binom{n_i}{l_i} p_i^{l_i}(1-p_i)^{n_i-l_i} \\ &= 0.9885. \end{aligned}$$

(Eryilmaz and Tank 2021) proposed a method for computing reliability of such system via matrix-geometric distributions. Using the methodology provided by (Eryilmaz and Tank 2021), the reliability of the consecutive 2-out-of-10:F system may also be computed from the following equation when $n_1 = 3$:

$$R(p_1,p_2) = \pi \mathbf{Q}^n (\mathbf{I} - \mathbf{Q})^{-1} \mathbf{u}'$$

where $\pi = (1,0,0,0,0)$,

$$\mathbf{Q} = \begin{bmatrix} p_2 & 0 & 0 & 0 & 1 \\ 0 & 0 & 0 & 0 & 0 \\ 0 & 1 & 0 & 0 & 0 \\ 0 & 0 & 1 & 0 & 0 \\ p_2(1-p_2) & 0 & 0 & 1 & 0 \end{bmatrix},$$

and

$$\mathbf{u}' = \begin{bmatrix} 0 \\ p_1(p_1-p_2)(p_1-2)(p_2-1) \\ (p_1-p_2)(1-p_1)(p_1p_2-p_2+1) \\ (p_1-p_2)(p_1-1)^2 \\ (1-p_1)^2 \end{bmatrix}. \quad \blacksquare$$

Series-parallel system is a system that consists of serially connected parallel subsystems and is useful for modeling production, supply chain systems, etc. Such a system has been extensively studied in the literature from different perspectives. In the following example, the reliability of discrete time series-parallel system is evaluated. For the details of this particular system model, see (Dembinska and Eryilmaz 2021).

Example 8.15. Let us consider a series-parallel system consisting of K parallel subsystems $S_1, S_2, \ldots, S_K$. The subsystem S_i is composed of n_i components with lifetimes $T_1^{(i)}, T_2^{(i)}, \ldots, T_{n_i}^{(i)}$. The components of the system work independently, so that $T_j^{(i)}$, $j = 1, 2, \ldots, n_i$, $i = 1, 2, \ldots, K$, are independent random variables. The components in the same subsystem are assumed to be identical, i.e. $T_1^{(i)}, T_2^{(i)}, \ldots, T_{n_i}^{(i)}$ are identically distributed for a fixed $i = 1, 2, \ldots, K$, while different subsystems may have different types of components. Let $F_i(t) = P\{T_j^{(i)} \leq t\}$ and $\bar{F}_i(t) = P\{T_j^{(i)} > t\}$ denote respectively the common cumulative distribution function and reliability function corresponding to $T_j^{(i)}$, $j = 1, 2, \ldots, n_i$. The random lifetimes $T_j^{(i)}$, $j = 1, 2, \ldots, n_i$, $i = 1, 2, \ldots, K$ are assumed to take on values only in the set $\{1, 2, \ldots\}$. Let $T_i = \max(T_1^{(i)}, T_2^{(i)}, \ldots, T_{n_i}^{(i)})$ denote the lifetime of subsystem S_i, $i = 1, 2, \ldots, K$, and T be the lifetime of the series-parallel system. Then, for $i = 1, 2, \ldots, K$ and $t = 1, 2, \ldots,$

$$\begin{aligned} P\{T_i \geq t\} &= P\{\max(T_1^{(i)}, T_2^{(i)}, \ldots, T_{n_i}^{(i)}) \geq t\} \\ &= 1 - P\{\max(T_1^{(i)}, T_2^{(i)}, \ldots, T_{n_i}^{(i)}) \leq t-1\} \\ &= 1 - F_i^{n_i}(t-1). \end{aligned}$$

Therefore, the mean time to failure of the series-parallel system can be computed from

$$\begin{aligned} E(T) &= \sum_{t=0}^{\infty} P\{T > t\} \\ &= \sum_{t=1}^{\infty} P\{T \geq t\} \\ &= \sum_{t=1}^{\infty} P\{\min(T_1, T_2, \ldots, T_K) \geq t\} \\ &= \sum_{t=1}^{\infty} \prod_{i=1}^{K} P\{T_i \geq t\} \\ &= \sum_{t=1}^{\infty} \prod_{i=1}^{K} \left(1 - F_i^{n_i}(t-1)\right). \end{aligned}$$

Example 8.16. For the series-parallel system described in Example 8.15, let the lifetime of the component j in subsystem i follow discrete phase-type distribution with representation $T_j^{(i)} \sim PH_{d_i}(\alpha^{(i)}, \mathbf{Q}^{(i)})$, for $j = 1,2,...,n_i$ and $i = 1,2,...,K$, where d_i is the order of the distribution, $\alpha^{(i)}$ is a row vector of dimension d_i and $\mathbf{Q}^{(i)}$ is a matrix of dimension $d_i \times d_i$. Using closure properties of phase-type distributions, Dembinska and Eryilmaz (2021) have shown that the the lifetime of the ith subsystem has a phase-type distribution with representation

$$T_i = \max(T_1^{(i)},...,T_{n_i}^{(i)}) \sim PH_{d_i^*}(\gamma_{n_i}^{(i)}, \mathbf{L}_{n_i}^{(i)}),$$

for $i = 1,2,...,K$, where $d_i^* = (d_i+1)^{n_i} - 1$,

$$\gamma_m^{(i)} = \left(\gamma_{m-1}^{(i)} \otimes \alpha^{(i)}, \mathbf{0}_{(d_i+1)^{m-1}+d_i-1}\right),$$

with $\mathbf{0}_{(d_i+1)^{m-1}+d_i-1} = (0,0,\ldots,0)_{1\times((d_i+1)^{m-1}+d_i-1)}$ denoting a row vector of zeros and

$$\mathbf{L}_m^{(i)} = \begin{bmatrix} \mathbf{L}_{m-1}^{(i)} \otimes \mathbf{Q}^{(i)} & \mathbf{L}_{m-1}^{(i)} \otimes \mathbf{u}^{(i)} & (\mathbf{I} - \mathbf{L}_{m-1}^{(i)})\mathbf{e}' \otimes \mathbf{Q}^{(i)} \\ \mathbf{0} & \mathbf{L}_{m-1}^{(i)} & \mathbf{0} \\ \mathbf{0} & \mathbf{0} & \mathbf{Q}^{(i)} \end{bmatrix}$$

for $m = 2,...,n_i$ with $\gamma_1^{(i)} = \alpha^{(i)}$ and $\mathbf{L}_1^{(i)} = \mathbf{Q}^{(i)}$. Because the minimum of phase-type random variables is also phase-type, the lifetime of the series-parallel system with K subsystems has a phase-type distribution with representation

$$T = \min(T_1,...,T_K) \sim PH_{d_T}(\beta_K, \mathbf{M}_K),$$

where $d_T = \prod_{i=1}^{K}((d_i+1)^{n_i} - 1)$, and

$$\beta_k = \beta_{k-1} \otimes \gamma_{n_k}^{(k)} \ , \ \mathbf{M}_k = \mathbf{M}_{k-1} \otimes \mathbf{L}_{n_k}^{(k)},$$

for $k = 2,...,K$ with $\beta_1 = \gamma_{n_1}^{(1)}$, $\mathbf{M}_1 = \mathbf{L}_{n_1}^{(1)}$. Thus, we can obtain the following equation to compute the mean time to failure of the system series-parallel system with K subsystems:

$$E(T) = \beta_K (\mathbf{I} - \mathbf{M}_K)^{-1} \mathbf{e}'. \quad \blacksquare$$

In a traditional setup of system reliability models, all components within the system are assumed to have the same contribution to the overall performance of the system. However, in most real-life situations, a system may consist of components which have different contributions. In a binary weighted-k-out-of-n system, component i has a weight of ω_i, $\omega_i > 0, i = 1,...,n$. The weight of a component actually represents the utility of the component. The binary weighted-k-out-of-n system works if and only if the total weight of working components exceeds a pre-specified level k (Wu and Chen 1994). Such systems are also known as threshold systems and they are widely available in real-life. For example, a wind power system that consists of a specified number of wind turbines can be modeled by a system with weighted components. In this case, the weight of a wind turbine corresponds to the power produced by the turbine. If p_i denotes the reliability of the ith component, then the reliability of the binary weighted-k-out-of-n system can be expressed as

$$R_k(p_1,...,p_n) = P\left\{\sum_{i=1}^{n}\omega_i X_i \geq k\right\}, \tag{8.7}$$

where X_i denotes the state of the ith component and $p_i = P\{X_i = 1\}, i = 1,...,n$.

The reliability given by (8.7) can be computed using various methods. In the following example, its calculation is illustrated using the method of universal generating function (UGF). For the details, see e.g. (Li and Zuo 2008).

Example 8.17. Consider a binary weighted-6-out-of-3 system consisting of $n = 3$ independent components with weights 3, 2, and 4. The system works if and only if the total weight of working components is at least $k = 6$. The UGFs for the three components are:

$$\begin{aligned}
U_1(z) &= (1-p_1)z^0 + p_1 z^3,\\
U_2(z) &= (1-p_2)z^0 + p_2 z^2,\\
U_3(z) &= (1-p_3)z^0 + p_3 z^4.
\end{aligned}$$

Then, the system's UGF is

$$\begin{aligned}
U_S(z) = {}&(1-p_1)(1-p_2)(1-p_3)z^0 + (1-p_1)(1-p_2)p_3 z^4\\
&+(1-p_1)p_2(1-p_3)z^2 + (1-p_1)p_2p_3z^6\\
&+p_1(1-p_2)(1-p_3)z^3 + p_1(1-p_2)p_3z^7\\
&+p_1p_2(1-p_3)z^5 + p_1p_2p_3z^9.
\end{aligned}$$

Because $k = 6$, summing the coefficients of the terms z^6, z^7, and z^9, the reliability of the weighted-6-out-of-3 system is found to be

$$\begin{aligned}
R_6(p_1,p_2,p_3) &= (1-p_1)p_2p_3 + p_1(1-p_2)p_3 + p_1p_2p_3\\
&= p_1p_3 + p_2p_3 - p_1p_2p_3. \quad \blacksquare
\end{aligned}$$

Consider the particular case when the weighted-k-out-of-n system consists of two types of components such that type I components have common weight ω and common reliability p_1 while type II components have common weight ω^* and common reliability p_2. Let m and $n-m$ denote respectively the number of components of type I and type II. Then, (Eryilmaz and Sarikaya 2014) presented the following equation for the reliability of such a weighted-k-out-of-n system:

$$R_k(p_1,p_2) = \sum\sum_{\substack{\omega y+\omega^* z\geq k\\ 0\leq y\leq m, 0\leq z\leq n-m}} \binom{m}{y} p_1^y(1-p_1)^{m-y}\binom{n-m}{z} p_2^z(1-p_2)^{n-m-z}.$$

This equation can be extended to the case when there are $K > 1$ different types of components.

As in weighted-k-out-of-n systems, the proper functioning of some systems depends on the overall generation/production capacity of system's components. Power generation system is a typical example for such a system. In the context of a power generation system, the reliability has two main aspects which are system adequacy aspect and system security aspect. The system adequacy aspect is concerned with the existence of sufficient facilities within the system to satisfy the consumer load demand while the system security aspect is the ability of the system to respond disturbances arising within the system (Billinton and Allan 1996). The latter one is in fact concerned with forced outage rates (or unreliabilities) of generating units within the system. One of the most commonly used reliability index for evaluating the reliability of a power system is the Loss of Load Expectation (LOLE) which is defined to be the expected period during which the load demand is greater than available power generation. If P^i_{sys} denotes the available power generation at time period i, then

$$LOLE = \sum_{i=1}^{N} P\left\{P^i_{sys} < L_i\right\},$$

where i is the equally sized time step (e.g. hour, day), N is the total number of time steps, and L_i is the load demand at time period i. If for example, we are interested in the expected number of hours in which the daily load will exceed the available capacity, then $N = 8736$ and hence the unit of LOLE is hr/year.

For a system whose capacity can be in one of m_{sys} states, the LOLE is computed by the following equation:

$$LOLE = \sum_{i=1}^{N} \sum_{j=1}^{m_{sys}} I\{b(j) < L_i\} P\left\{P_{sys} = b(j)\right\}, \tag{8.8}$$

where m_{sys} is the total number of states for the combined system, $b(1), b(2), ..., b(m_{sys})$ are the possible production capacities of the system, and $I(A) = 1$ if A occurs and $I(A) = 0$, otherwise.

Example 8.18. Consider the power system in Example 1.1. Recall that the system has three generating units with capacities 20 MW, 10 MW, and 30 MW. At any instant of time, the system may produce 0 MW, 10 MW, 20 MW, 30 MW, 40 MW, 50 MW, or 60 MW powers depending on the available units. In Example 1.1, the following probabilities have been computed for the total generation capacity of the system:

$$\begin{aligned}
P\left\{P_{sys} = 0\right\} &= 0.0001 \\
P\left\{P_{sys} = 10\right\} &= 0.0034 \\
P\left\{P_{sys} = 20\right\} &= 0.0020 \\
P\left\{P_{sys} = 30\right\} &= 0.0659 \\
P\left\{P_{sys} = 40\right\} &= 0.0451 \\
P\left\{P_{sys} = 50\right\} &= 0.0265 \\
P\left\{P_{sys} = 60\right\} &= 0.8570.
\end{aligned}$$

Assume that the load for the half of the day is constant and 20 MW and it is 10 MW for the other half of the day. Then, using (8.8), the LOLE is found to be

$$
\begin{aligned}
LOLE &= \sum_{i=1}^{24}\sum_{j=1}^{7} I\{b(j) < L_i\} P\{P_{sys} = b(j)\} \\
&= 12P\{P_{sys} = 0\} + 12\left[P\{P_{sys} = 0\} + P\{P_{sys} = 10\}\right] \\
&= 0.0432 \text{ hours/day. } \blacksquare
\end{aligned}
$$

In reliability analysis, systems/components can be classified with respect to their repair status as repairable or nonrepairable. In the repairable case, the repair time is mostly modeled as a random variable and a proper probability distribution is fitted. The probabilistic analysis of repairable systems/components is relatively more difficult when it is compared with nonrepairable systems/components. This is mainly due to the inclusion of repair time distribution and repair policy in the analysis. In the following example, a discrete time repairable system that includes one active and one standby component is analyzed by using phase-type distributions.

Example 8.19. Consider the system model that has been previously considered in Example 1.6. Recall that the system consists of one active and one cold standby component. The lifetime of the system is

$$
T = \sum_{i=0}^{M} X_i,
$$

where $\{M = m\}$ iff $\{X_1 \geq R_1, ..., X_{m-1} \geq R_{m-1}, X_m < R_m\}$, and $X_0, X_1, X_2, ...$ denote successive lifetimes of the active component, and R_i represents the repair time of a component upon the ith failure, $i = 1, 2, ...$ (Kan and Eryilmaz 2021) has obtained the following expression for the probability generating function of the system's lifetime when $X_i \sim PH_{m_1}(\mathbf{a}, \mathbf{Q}_1), i = 0, 1, 2, ...$ and $R_i \sim PH_{m_2}(\mathbf{b}, \mathbf{Q}_2), i = 1, 2, ...$

$$
\phi_T(z) = \frac{\phi(z) z (\mathbf{a} \otimes \mathbf{b}) \left(\mathbf{I}_{m_1 \times m_2} - z(\mathbf{Q}_1 \otimes \mathbf{Q}_2)\right)^{-1} (\mathbf{u}' \otimes \mathbf{Q}_2 \mathbf{e}')}{1 - \phi(z) + z(\mathbf{a} \otimes \mathbf{b}) \left(\mathbf{I}_{m_1 \times m_2} - z(\mathbf{Q}_1 \otimes \mathbf{Q}_2)\right)^{-1} (\mathbf{u}' \otimes \mathbf{Q}_2 \mathbf{e}')} \tag{8.9}
$$

where $\phi(z) = E(z^{X_1}) = \mathbf{a} z (\mathbf{I}_{m_1} - z\mathbf{Q}_1)^{-1} \mathbf{u}'$, and $\mathbf{I}_m$ is $m \times m$ identity matrix. In the special case when $P\{X_i = x\} = p_1(1 - p_1)^{x-1}, i = 0, 1, 2, ...$ and $P\{R_i = x\} = p_2(1 - p_2)^{x-1}, i = 1, 2, ...; x = 1, 2, ...$ From (8.9), the probability generating of the system's lifetime is (Kan and Eryilmaz 2020)

$$
\phi_T(z) = \frac{a_1 z + a_2 z^2}{1 + b_1 z + b_2 z^2},
$$

where $a_1 = 0, a_2 = p_1^2(1 - p_2)$,

$$
b_1 = 2(p_1 - p_1 p_2 - 1) + p_2,
$$

$$
b_2 = 1 - 2p_1 - p_2 + p_1^2 - p_1^2 p_2 + 2p_1 p_2 = (1 - p_1)^2 (1 - p_2).
$$

Thus, the system's lifetime has a phase-type distribution. The following recursive relation for the probability mass function of T can be obtained using (4.6):

$$\begin{aligned} P\{T=1\} &= 0, \\ P\{T=2\} &= p_1^2(1-p_2), \\ P\{T=n\} &= -b_1 P\{T=n-1\} - b_2 P\{T=n-2\} \text{ if } n>2. \end{aligned} \tag{8.10}$$

The set of recursive equations given by (8.10) is an alternative to the combinatorial equation (1.6). For a numerical illustration, let $p_1 = 0.1$ and $p_2 = 0.2$. Then, from (8.10)

$$\begin{aligned} P\{T=2\} &= 0.0080, \\ P\{T=3\} &= 0.0131, \\ P\{T=4\} &= 0.0163, \\ P\{T=5\} &= 0.0183, \\ P\{T=6\} &= 0.0194, \\ P\{T=20\} &= 0.0162. \end{aligned}$$

For the same values of the parameters p_1 and p_2, the mean time to failure of the system is found to be

$$E(T) = 45.$$

8.2 SHOCK MODELS

Shocks models have attracted a great deal of attention in reliability engineering and applied probability. Numerous shock models have been defined and studied from different perspectives. In a shock model, a component/system is subject to shocks of random magnitudes at random times and the component/system fails if a certain rule corresponding to the magnitudes of shocks, and/or times between successive shocks is observed. Thus, in general, the lifetime of the component/system can be represented as a compound random variable which becomes a function of random magnitudes of shocks and the times between successive shocks. A shock can be viewed as a load/stress which may refer to mechanical stress, a voltage or internally generated stress such as temperature.

For a system that is subject to shocks over time, let X_i denote the time between ith and $(i+1)$th shocks for $i = 1,2,...$ and X_0 be the time until the first shock. The random variables $X_0, X_1, ...$ are assumed to have a discrete distribution with common probability mass function $P\{X_i = x\}, i = 0,1,...; x = 1,2,....$ For a stopping random variable M, we focus on the shock models in which the lifetime of the system has the following representation:

$$T = \sum_{i=0}^{M} X_i.$$

The random variable M in fact represents the number of shocks until system failure.

In the following, the reliability of the system is examined under well-known shock models when the times between shocks follow discrete probability distribution. The continuous time versions of these models have been investigated by many authors. See, e.g. (Gut and Hüsler 1999), (Mallor and Omey 2001), (Wang and Zhang 2005), (Li and Kong 2007), and (Parvardeh and Balakrishnan 2015). The random variable Y_i denotes the magnitude of the ith shock, $i = 1, 2, \ldots$. The sequence of independent and identically distributed random variables $Y_1, Y_2, \ldots$ is assumed to be independent of $X_0, X_1, \ldots$.

Example 8.20. According to the extreme shock model, the system breaks down due to a single shock that is above a critical level. Let c be a critical shock threshold for the failure of the system. Then, under this model, the lifetime of the system is defined by

$$T = \sum_{i=0}^{M} X_i, \tag{8.11}$$

where $\{M = 0\}$ iff $\{Y_1 > c\}$ and

$$\{M = m\} \text{ iff } \{Y_1 \leq c, \ldots, Y_m \leq c, Y_{m+1} > c\},$$

for $m > 0$. (Eryilmaz and Kan 2021a) obtained the following equation for the probability generating function of the system's lifetime defined by (8.11) when $X_i \sim PH_m(\mathbf{a}, \mathbf{Q})$ and $r = P\{Y_{i+1} \leq c\}, i = 0, 1, 2, \ldots$.

$$\phi_T(z) = \frac{\phi(z)(1-r)}{1-\phi(z)r}, \tag{8.12}$$

where $\phi(z) = E(z^{X_1}) = \mathbf{a}z(\mathbf{I} - z\mathbf{Q})^{-1}\mathbf{u}'$. Differentiating (8.12) with respect to z, one obtains

$$\phi_T'(z) = \frac{(1-r)\phi'(z)}{(1-\phi(z)r)^2}.$$

Thus,

$$E(T) = \phi_T'(1) = \frac{\phi'(1)}{1-r} = \frac{E(X)}{1-r}.$$

In particular, if $P\{X_i = x\} = p(1-p)^{x-1}, x = 1, 2, \ldots$, i.e. the times between successive shocks have geometric distribution with mean $\frac{1}{p}$, then

$$\phi_T(z) = \frac{p(1-r)z}{1 - z(1 - p(1-r))}.$$

That is, the system's lifetime has geometric distribution with mean $\frac{1}{p(1-r)}$ (Eryilmaz and Kan 2021).

Example 8.21. For $\delta > 1$, assume that the system fails when the time between two consecutive shocks falls below a fixed threshold δ. This model is known to be δ-shock model. In this model, the magnitudes of the shocks are not taken into account and the system failure only depends on the lengths of the intershock times. The lifetime of the system is defined by

$$T = \sum_{i=0}^{M} X_i, \tag{8.13}$$

where $\{M = m\}$ iff $\{X_1 \geq \delta, ..., X_{m-1} \geq \delta, X_m < \delta\}$. In this case, the random variable M which denotes the number of shocks until the failure of the system depends on the random variables $X_0, X_1, ...$ (Eryilmaz and Kan 2021a) has obtained the following equation for the probability generating function of the random variable (8.13) when $X_i \sim PH_m(\mathbf{a}, \mathbf{Q}), i = 0, 1, 2, ...$:

$$\phi_T(z) = \frac{\phi(z)\,[\phi(z) - A_\delta(z)]}{1 - A_\delta(z)}, \tag{8.14}$$

where

$$A_\delta(z) = \mathbf{a}z(\mathbf{I} - \mathbf{Q})(\mathbf{I} - z\mathbf{Q})^{-1}(z\mathbf{Q})^{\delta-1}\mathbf{e}',$$

and

$$\phi(z) = E(z^{X_1}) = \mathbf{a}z(\mathbf{I} - z\mathbf{Q})^{-1}\mathbf{u}'.$$

In particular, if $P\{X_i = x\} = p(1-p)^{x-1}$, $x = 1, 2, ...,$ then we have the following representation for (8.14):

$$\phi_T(z) = \frac{p^2z^2 - p^2(1-p)^{\delta-1}z^{\delta+1}}{1 - 2(1-p)z + (1-p)^2z^2 - p(1-p)^{\delta-1}z^\delta + p(1-p)^\delta z^{\delta+1}}, \tag{8.15}$$

for $\delta > 1$. The probability generating function given by (8.15) has the same form with (5.2) and hence if the times between successive shocks have geometric distribution, then the system's lifetime has matrix-geometric distribution with representation $T \sim MG(\pi, \mathbf{Q}, \mathbf{u})$. Using (5.3), the survival function of T can be expressed as

$$P\{T > t\} = \pi\mathbf{Q}^t(\mathbf{I} - \mathbf{Q})^{-1}\mathbf{u}', \tag{8.16}$$

where $\pi = (1, 0, ..., 0)$, the matrix $\mathbf{Q}$ with dimension $(\delta + 1) \times (\delta + 1)$ and the vector $\mathbf{u}'$ with dimension $(\delta + 1) \times 1$ are given respectively by

$$\mathbf{Q} = \begin{bmatrix} 2(1-p) & 0 & 0 & \cdots & 0 & 1 \\ -p(1-p)^\delta & 0 & 0 & \cdots & 0 & 0 \\ p(1-p)^{\delta-1} & 1 & 0 & \cdots & 0 & 0 \\ 0 & 0 & 1 & & 0 & 0 \\ \vdots & \vdots & \vdots & \ddots & \vdots & \vdots \\ -(1-p)^2 & 0 & 0 & & 1 & 0 \end{bmatrix},$$

and

$$\mathbf{u}' = \begin{bmatrix} 0 \\ -p^2(1-p)^{\delta-1} \\ 0 \\ \vdots \\ 0 \\ p^2 \end{bmatrix}.$$

For an illustration, in Table 8.4, we compute survival probabilities using (8.16) when $\delta = 3$ for selected values of the parameter p. As it is clear from the Table 8.4, the survival probability is a decreasing function of the parameter p. That is, a decrease in mean time between shocks leads to a decrease in survival probability.

Table 8.4
Survival Probabilities for the System When $\delta = 3$

t	$p = 0.1$	$p = 0.2$
3	0.9720	0.8960
5	0.9404	0.7987
10	0.8662	0.5977
15	0.7979	0.4474
20	0.7350	0.3349

Example 8.22. Consider the case whenever the system fails if either the time between two successive shocks falls below a fixed threshold δ or the magnitude of a single shock exceeds the level c. This model is known to be a mixed shock model and combines extreme and δ-shock models. The system's lifetime is defined as

$$T = \sum_{i=0}^{M} X_i, \tag{8.17}$$

where $\{M = 0\}$ iff $\{Y_1 > c\}$ and $\{M = m\}$ iff $\{Y_1 \le c, X_1 \ge \delta \wedge Y_2 \le c, ..., X_{m-1} \ge \delta \wedge Y_m \le c, X_m < \delta \vee Y_{m+1} > c\}$ for $m > 0$. Let $X_i \sim PH_m(\mathbf{a}, \mathbf{Q}), i = 0,1,2,...,$ $s = P\{X_i \ge \delta\}, i = 0,1,2,...,$ and $r = P\{Y_i \le c\}, i = 1,2,....$ Then, the probability generating function of T is (Eryilmaz and Kan 2021a):

$$\phi_T(z) = \frac{(1-r)\phi(z)\left[1 - A_\delta(z)r\right] + r\phi(z)\left[\phi(z) - A_\delta(z)r\right]}{1 - A_\delta(z)r}, \tag{8.18}$$

for $\delta > 1$, where

$$A_\delta(z) = \mathbf{a}z(\mathbf{I}-\mathbf{Q})(\mathbf{I}-z\mathbf{Q})^{-1}(z\mathbf{Q})^{\delta-1}\mathbf{e}',$$

and

$$\phi(z) = E(z^{X_1}) = \mathbf{a}z(\mathbf{I}-z\mathbf{Q})^{-1}\mathbf{u}'.$$

For an illustration, let $P\{X_i = x\} = p(1-p)^{x-1}$, $x = 1,2,...$ for $i = 0,1,2,...$ and $\delta = 3$. Then, from (8.18), the probability generating function of the random variable defined by (8.17) is found to be

$$\phi_T(z) = \frac{c_1 z + c_2 z^2 + c_3 z^3 + c_4 z^4}{1 + b_1 z + b_2 z^2 + b_3 z^3 + b_4 z^4}, \tag{8.19}$$

where

$$c_1 = p(1-r),\ c_2 = p(p+r-1),\ c_3 = 0,\ c_4 = -rp^2(1-p)^2,$$

and

$$b_1 = -2(1-p),\ b_2 = (1-p)^2,\ b_3 = -rp(1-p)^2,\ b_4 = rp(1-p)^3.$$

The probability generating function is rational and has the same form with (5.2). Thus, the system's lifetime has matrix-geometric distribution with representation $T \sim MG(\pi, \mathbf{Q}, \mathbf{u})$, where $\pi = (1,0,0,0)$, and

$$\mathbf{Q} = \begin{bmatrix} 2(1-p) & 0 & 0 & 1 \\ -rp(1-p)^3 & 0 & 0 & 0 \\ rp(1-p)^2 & 1 & 0 & 0 \\ -(1-p)^2 & 0 & 1 & 0 \end{bmatrix},$$

and

$$\mathbf{u}' = \begin{bmatrix} p(1-r) \\ -rp^2(1-p)^2 \\ 0 \\ p(p+r-1) \end{bmatrix}.$$

For an illustration, let $p = 0.1$ and $r = 0.9$. Then,

$$P\{T > 3\} = 0.9477,$$
$$P\{T > 5\} = 0.9034,$$
$$P\{T > 10\} = 0.8020,$$
$$P\{T > 15\} = 0.7119. \quad \blacksquare$$

(Eryilmaz and Kan 2021b) modified the classical extreme shock model in the following way. The system is subject to shocks which have random magnitudes denoted by $Y_1, Y_2, \ldots$. The distribution of the magnitudes of shocks changes after the first shock of size above d_1. That is,

$$Y_1, \ldots, Y_{N_{d_1}} \sim G_1,$$

and

$$Y_{N_{d_1}+1}, Y_{N_{d_1}+2}, \ldots \sim G_2,$$

where N_{d_1} denotes the number of shocks until the first shock of size above d_1. The system fails upon the occurrence of the first shock of size greater than d_2 ($d_2 > d_1$). As noted by (Eryilmaz and Kan 2021b), the shock $Y_{N_{d_1}+1}$ actually triggers the change in the shock size distribution. Such a shock can be called as the trigger shock. Define

$$p_1 = P\{Y_i \le d_1\} = G_1(d_1) \text{ for } i = 1, \ldots, N_{d_1},$$
$$p_2 = P\{Y_i \le d_2\} = G_1(d_2) \text{ for } i = 1, \ldots, N_{d_1},$$

and

$$p_3 = P\{Y_i \leq d_2\} = G_2(d_2) \text{ for } i = N_{d_1}+1, \ldots$$

If N_{d_2} denotes the number of shocks until the first shock of size above d_2, then (Eryilmaz and Kan 2021b) obtained the following expression for the joint probability mass function of (N_{d_1}, N_{d_2}) for $d_1 < d_2$:

$$P\{N_{d_1} = m, N_{d_2} = n\}$$
$$= \begin{cases} p_1^{m-1}(p_2 - p_1)p_3^{n-m-1}(1-p_3), & \text{if } n > m \\ p_1^{n-1}(1-p_2), & \text{if } n = m. \end{cases}$$

The marginal probability mass function of N_{d_2} was also obtained by (Eryilmaz and Kan 2021b):

$$P\{N_{d_2} = n\} = \left(\frac{p_3 - p_2}{p_3 - p_1}\right) p_1^{n-1}(1-p_1) + \left(\frac{p_2 - p_1}{p_3 - p_1}\right) p_3^{n-1}(1-p_3),$$

for $p_1 \neq p_3$, and

$$P\{N_{d_2} = n\} = \left(\frac{p_2 - p_3}{1 - p_3}\right)(n-1)p_3^{n-2}(1-p_3)^2 + \left(\frac{1-p_2}{1-p_3}\right) p_3^{n-1}(1-p_3),$$

for $p_1 = p_3$.

If $X_1, X_2, \ldots$ represent the times between successive shocks, then the lifetime of the system is

$$T = \sum_{i=1}^{N_{d_2}} X_i.$$

The probability generating function of N_{d_2} is obtained as

$$E(z^{N_{d_2}}) = \frac{(1-p_2)z + (p_2 - p_1 - p_3 + p_1 p_3)z^2}{1 - (p_1 + p_3)z + p_1 p_3 z^2}.$$

The random variable N_{d_2} can be represented by a matrix-geometric distribution. Indeed, considering its probability generating function, one obtains $N_{d_2} \sim MG(\pi, \mathbf{Q}, \mathbf{u})$, where $\pi = (1, 0)$,

$$\mathbf{Q} = \begin{bmatrix} -d_1 & 1 \\ -d_2 & 0 \end{bmatrix}, \quad \mathbf{u}' = \begin{bmatrix} c_1 \\ c_2 \end{bmatrix},$$

where

$$c_1 = 1 - p_2, \quad c_2 = p_2 - p_1 - p_3 + p_1 p_3,$$

and

$$d_1 = -(p_1 + p_3), \quad d_2 = p_1 p_3.$$

Thus, if $X_1, X_2, \ldots$ is a sequence of independent and identically distributed random variables with common discrete distribution and independent of shock magnitudes

$Y_1, Y_2, \ldots$, then the probability generating function of the system's lifetime is obtained as

$$E(z^T) = \frac{(1-p_2)\phi(z) + (p_2 - p_1 - p_3 + p_1 p_3)(\phi(z))^2}{1-(p_1+p_3)\phi(z) + p_1 p_3(\phi(z))^2},$$

where $\phi(z) = E(z^X)$ is the common probability generating function of $X_1, X_2, \ldots$

8.3 AGE REPLACEMENT POLICIES

Age replacement policies are of special interest in reliability engineering and operational research. Numerous age replacement policies have been defined and studied from different perspectives. In a classical age replacement model, two costs appear: The cost of unplanned replacement and the cost of scheduled replacement. The former one is in fact the cost at system failure while the latter is the replacement cost before failure. Manifestly, the cost of scheduled (preventive) replacement is smaller than the cost at system failure since additional costs appear upon system failure. The main optimization problem is to find the optimal time point that minimizes the expected cost per unit of time. Clearly, such an optimization problem heavily depends on the lifetime distribution of the unit/system.

Below, the mathematical formulation of the problem is presented for the discrete time age replacement policy. The system is assumed to have a discrete lifetime distribution and it is replaced preventively at time n $(n = 1, 2, \ldots)$ or correctively at system failure, whichever occurs first. Let c_1 denote the replacement cost at failure and c_2 denote the replacement cost at time n before failure, i.e. when the nth operation cycle is completed, where $c_1 > c_2$. Then, for a system with lifetime T, the expected cost is

$$c_1 P\{T \leq n\} + c_2 P\{T > n\} = c_1 + (c_2 - c_1)P\{T > n\}.$$

Thus, the expected cost rate per unit of time is

$$C(n) = \frac{c_1 + (c_2 - c_1)P\{T > n\}}{E(\min(T,n))}. \tag{8.20}$$

The problem is to find n^*, i.e. the number of operation cycles to preventive replace that minimizes the expected cost rate defined by (8.20). (Nakagawa and Osaki 1977) obtained the following conditions for the optimal n^*. Let $r(n)$ be the hazard rate function corresponding to lifetime random variable T, i.e.

$$r(n) = \frac{P\{T = n\}}{P\{T \geq n\}}.$$

Assume that $r(n)$ is monotonically increasing. If

$$\lim_{n \longrightarrow \infty} r(n) > \frac{1}{E(T)} \frac{c_1}{c_1 - c_2},$$

then there exists a finite and unique n^* that satisfies

$$L(n) \geq \frac{c_2}{c_1 - c_2} \text{ and } L(n-1) < \frac{c_2}{c_1 - c_2},$$

where

$$L(n) = \begin{cases} 0, & \text{if } n = 0 \\ r(n+1)\sum\limits_{i=1}^{n}\sum\limits_{j=i}^{\infty} P\{T=j\} - \sum\limits_{j=1}^{n} P\{T=j\}, & \text{if } n = 1,2,... \end{cases}.$$

If

$$\lim_{n \longrightarrow \infty} r(n) \le \frac{1}{E(T)} \frac{c_1}{c_1 - c_2},$$

then $n^* = \infty$, i.e. no preventive replacement.

Combined continuous and discrete age-replacement policies have been considered by (Nakagawa 1985). In this case, a unit is replaced before failure at a specific time of age or at a certain number of uses, whichever occurs first.

(Eryilmaz 2021) studied the above mentioned discrete time age replacement policy when the system's lifetime has discrete phase-type distribution. Let $T \sim PH_m(\pi, \mathbf{Q})$. Then, as shown by (Eryilmaz 2021), the expected cost rate per unit of time defined by (8.20) becomes

$$C(n) = \frac{c_1 + (c_2 - c_1)\pi \mathbf{Q}^n \mathbf{e}'}{\pi(\mathbf{I} - \mathbf{Q}^n)(\mathbf{I} - \mathbf{Q})^{-1}\mathbf{e}'}. \tag{8.21}$$

For the phase-type lifetime distribution, the conditions for the optimal n^* can be reformulated as follows. Let the hazard rate function

$$r(n) = \frac{P\{T = n\}}{P\{T \ge n\}} = \frac{\pi \mathbf{Q}^{n-1}\mathbf{u}'}{\pi \mathbf{Q}^{n-1}\mathbf{e}'}$$

be monotonically increasing. If

$$\lim_{n \longrightarrow \infty} r(n) > \frac{1}{E(T)} \frac{c_1}{c_1 - c_2} = \frac{1}{\pi(\mathbf{I} - \mathbf{Q})^{-1}\mathbf{e}'} \frac{c_1}{c_1 - c_2},$$

then there exists a finite and unique n^* which satisfies

$$L(n) \ge \frac{c_2}{c_1 - c_2} \text{ and } L(n-1) < \frac{c_2}{c_1 - c_2},$$

where

$$L(n) = \frac{\pi \mathbf{Q}^n \mathbf{u}'}{\pi \mathbf{Q}^n \mathbf{e}'} \pi(\mathbf{I} - \mathbf{Q}^n)(\mathbf{I} - \mathbf{Q})^{-1}\mathbf{e}' + \pi \mathbf{Q}^n \mathbf{e}' - 1.$$

If

$$\lim_{n \longrightarrow \infty} r(n) \le \frac{1}{\pi(\mathbf{I} - \mathbf{Q})^{-1}\mathbf{e}'} \frac{c_1}{c_1 - c_2},$$

then $n^* = \infty$, i.e. no preventive replacement.

As it is clear from the above inequalities, the existence of a finite and unique n^* depends on the monotonicity and the limiting point of the hazard rate. (Eryilmaz 2021) has obtained necessary conditions for the existence of the optimal n^* when the system's lifetime has discrete phase-type distribution of order m. The following example gives the necessary conditions when the system's lifetime has phase-type distribution of order $m=2$. As it is clear from the Example, the existence of finite and unique n^* can be checked via the coefficients in the probability generating function of T.

Example 8.23. Let $T \sim PH_2(\pi, \mathbf{Q})$ with the probability generating function

$$\phi_T(z) = \pi_0 + \frac{a_1 z + a_2 z^2}{1 + b_1 z + b_2 z^2},$$

where $\pi_0 = 1 - \pi\mathbf{e}'$. If $b_1^2 - 4b_2 > 0$, $b_2 > 0$, and

$$1 - \left[\frac{-b_1 + \sqrt{b_1^2 - 4b_2}}{2}\right] > \frac{1 + b_1 + b_2}{a_1 - b_1 + 2(a_2 - b_2)} \frac{c_1}{c_1 - c_2}, \tag{8.22}$$

then there exists a finite and unique n^* that minimizes (8.21) (Eryilmaz 2021).

Example 8.24. Consider the cold standby repairable system described in Example 8.18. The system's lifetime has a phase-type distribution of order $m=2$. Let $p_1 = 0.1$ and $p_2 = 0.15$. Then, $b_1^2 - 4b_2 = 0.0684$, $b_2 = 0.6885$. Suppose that $c_1 = 12$ and $c_2 = 1$. In this case, (8.22) is satisfied and hence there exists a finite and unique n^* that minimizes the expected cost rate per unit of time. The value of n^* is found to be $n^* = 12$. On the other hand, if $c_1 = 10$ and $c_2 = 1$, then $n^* = \infty$, i.e. no scheduled replacement should be made. This example clearly exhibits the importance of the cost values on optimal replacement time. ■

Chien (2012) investigated the effects of a renewing free-replacement warranty on the discrete age replacement policy. He developed cost models from the customer's perspective and compared optimal replacement times for both warranted and non-warranted products when the failure time distribution of the product under concern is discrete. Let W denote warranty period in discrete time. The warranty period means the number of operation cycles that covered by the warranty. In the case of a product failure, a failure replacement is performed with a downtime cost C_d, and a purchasing cost C_p. If a planned replacement is made, then only the cost C_p is incurred. If the warranty is not involved in the model, then the policy is same with the classical discrete age replacement policy whose corresponding long-run expected cost rate is (8.20) with $c_1 = C_d + C_p$ and $c_2 = C_p$. That is, we have the following mean cost rate under the cost model without warranty:

$$C(n) = \frac{C_p + C_d \sum\limits_{j=1}^{n} p_j}{\sum\limits_{j=1}^{n} \sum\limits_{i=j}^{\infty} p_i},$$

where $p_j = P\{T = j\}$ is the probability that the product fails at operation cycle j, $j = 1, 2, \ldots$.

For the cost model with warranty, as stated by Chien (2012), the two cases appear: $n > W$ and $n \leq W$. In the first case when $n > W$, the operation cycle for preventive replacement is scheduled after the warranty period. For $n > W$, the expected cost is

$$C_d P\{T \le W\} + (C_d + C_p) P\{W < T \le n\} + C_p P\{T > n\}.$$

Thus, the expected cost rate per unit of time is

$$C_1(n) = \frac{C_p \sum_{j=W+1}^{\infty} p_j + C_d \sum_{j=1}^{n} p_j}{\sum_{j=1}^{n} \sum_{i=j}^{\infty} p_i}.$$

For the case $n \le W$, the operation cycle for preventive replacement is scheduled within the warranty. In this case, the expected cost becomes

$$C_d P\{T \le n\} + C_p P\{T > n\}.$$

Therefore, the expected cost rate per unit of time is

$$C_2(n) = \frac{C_p \sum_{j=n+1}^{\infty} p_j + C_d \sum_{j=1}^{n} p_j}{\sum_{j=1}^{n} \sum_{i=j}^{\infty} p_i}.$$

Let n_i^* denote the optimal age that minimizes $C_i(n), i = 1, 2$. Chien (2012) obtained the following results for the global optimal replacement cycle n_W^* when the warranty is included. Let

$$H(n) = r(n+1) \sum_{j=1}^{n} \sum_{i=j}^{\infty} p_i - \sum_{j=1}^{n} p_j.$$

Then,

(i) For $C_d \le C_p$, if $H(W) < \frac{C_p}{C_d} \sum_{j=W+1}^{\infty} p_j$, then $n_W^* = n_1^* > W$; otherwise $n_W^* = W$.

(ii) For $C_d > C_p$,

(a) if $H(W) < \frac{C_p}{C_d} \sum_{j=W+1}^{\infty} p_j$, then $n_W^* = n_1^* > W$;

(b) if $\frac{C_p}{C_d} \sum_{j=W+1}^{\infty} p_j \le H(W) \le \frac{C_p}{(C_d - C_p)}$, then $n_W^* = W$;

(c) if $H(W) > \frac{C_p}{(C_d - C_p)}$, then $n_W^* = n_2^* < W$.

8.4 QUALITY CONTROL

Statistical quality control is concerned with monitoring, controlling and improving the quality of products and manufacturing processes. It effectively uses probabilistic and statistical methods. Control charts are key elements in statistical quality control. They are used to monitor the manufacturing process for quality improvement.

Generally speaking, a production line or manufacturing process can be classified as in-control or out-of-control. When a process operates only under chance causes, it is said to be in-control. That is, if the process is in-control, then it is stable with common causes. In the event there is a change in the process, a good control chart should detect it as quickly as possible and give an out-of-control signal. The quicker the detection of the correct signal, the more efficient the chart (Chakraborti and Eryilmaz 2007).

In statistical process control, the number of samples/subgroups that needs to be collected before the first out-of-control signal that is given by a control chart is called a run-length random variable. The expected value of the run-length random variable is called Average Run Length (ARL) in statistical process control. It has been effectively used as a performance indicator of a control chart when the sampling interval remains constant. That is, ARL can be used to compare the performances of two different control charts for the same manufacturing process.

The computation of ARL for a given control chart is closely related to the distribution of a waiting time random variable which is specified by the run rule. Thus, Markov chain approach and in particular phase-type distributions can be effectively used to compute run-length distribution and ARL after defining the underlying Markov chain.

Example 8.25. Let X be a random variable that measures the quality of the product. The independent sample means $\bar{X}_1, \bar{X}_2, \ldots$ corresponding to subsamples are successively observed, where $\bar{X}_i \sim N(\mu, \sigma^2/n), i = 1, 2, \ldots$, where σ^2 is known. According to the Shewart control chart, the out-of-control is given at the first stage n such that $|Z_n| > c$, where $Z_i = (\bar{X}_i - \mu_0)/(\sigma/\sqrt{n})$. The value of c is chosen to be three usually. Thus, for this well-known Shewart control chart, the run length random variable T is defined by

$$\{T = n\} \Longleftrightarrow \{|Z_1| \leq c, \ldots, |Z_{n-1}| \leq c, |Z_n| > c\},$$

for $n = 1, 2, \ldots$. That is, the run-length random variable has a geometric distribution with mean $\frac{1}{P\{|Z_n| > c\}}$. Thus, the ARL corresponding to the chart is

$$ARL = \frac{1}{P\{|Z_n| > c\}}. \quad \blacksquare$$

The procedure given by Example 8.25 is known to have a disadvantage since it is not sensitive to small shifts in the mean. To improve the chart's sensitivity to small changes in the mean, supplementary runs rules have been suggested by (Champ and Woodall 1987). The rules can be defined as follows: An out-of-control signal is given if k of the last m standardized sample means fall in the interval (a, b), where $k \leq m$ and $a < b$. This rule will be denoted by $R(k, m, a, b)$. Clearly, the rule given by Example 8.19 is $\{R(1, 1, -\infty, -3), R(1, 1, 3, \infty)\}$ when $c = 3$.(Champ and Woodall 1987) considered various runs rules and their combinations and computed ARL values to increase the sensitivity of the Shewhart chart to small shifts in the mean. In the following example, for an illustration, we compute the ARL for the combination of the rules defined by Rule 1 and Rule 5 in (Champ and Woodall 1987). Our method is based on the use of discrete phase-type distributions.

Example 8.26. The corresponding rules are given by

$$C_1 = \{R(1,1,-\infty,-3), R(1,1,3,\infty)\},$$
$$C_5 = \{R(2,2,-3,-2), R(2,2,2,3)\}.$$

We are interested in the ARL of the combined chart defined by $C_1 \cup C_5$. In this case, the out-of-control signal is given if a single standardized sample mean falls in $(-\infty,-3)$ or $(3,\infty)$ or two consecutive standardized sample means fall in $(-3,-2)$ or two consecutive standardized sample means fall in $(2,3)$. Thus, a Markov chain with seven states can be used to compute the ARL of the chart $C_1 \cup C_5$. Note that in the following, a point means a standardized sample mean $Z_i = (\bar{X}_i - \mu_0)/(\sigma/\sqrt{n})$.

State 0: A point is in $(-2,2)$
State 1: A point is in $(-\infty,-3)$
State 2: A point is in $(3,\infty)$
State 3: A point is in $(-3,-2)$
State 4: A point is in $(2,3)$
State 5: Two consecutive points are in $(-3,-2)$
State 6: Two consecutive points are in $(2,3)$

The absorbing states of the corresponding Markov chain are $\{1\},\{2\},\{5\},\{6\}$. Thus, the run-length random variable has a phase-type distribution with representation $T \sim PH_3(\pi,\mathbf{Q})$ with $\pi = (1,0,0)$, and

$$\mathbf{Q} = \begin{bmatrix} p_0 & p_3 & p_4 \\ p_0 & 0 & p_4 \\ p_0 & p_3 & 0 \end{bmatrix}.$$

The probability mass function of the run-length random variable is

$$P\{T=n\} = (1,0,0) \begin{bmatrix} p_0 & p_3 & p_4 \\ p_0 & 0 & p_4 \\ p_0 & p_3 & 0 \end{bmatrix}^{n-1} \begin{bmatrix} p_1+p_2 \\ p_1+p_2+p_3 \\ p_1+p_2+p_4 \end{bmatrix},$$

where

$$p_1 = P\{Z < -3-d\},\ p_2 = P\{Z > 3-d\},$$
$$p_3 = P\{-3-d < Z < -2-d\},\ p_4 = P\{2-d < Z < 3-d\},$$
$$p_0 = 1 - p_1 - p_2 - p_3 - p_4,$$

$d = \sqrt{n}\frac{|\mu-\mu_0|}{\sigma}$ and Z is standard normal random variable. Thus, the ARL for the chart $C_1 \cup C_5$ can be computed from

$$ARL = E(T) = \pi(\mathbf{I}-\mathbf{Q})^{-1}\mathbf{e}'.$$

The value of d is chosen to be zero to compute the in-control ARL. In Table 8.5, we compute ARL values of the charts C_1 and $C_1 \cup C_5$ for various values of the shift d.

Note that the ARL for the chart C_1 is $\frac{1}{p_1+p_2}$. As it is clear from Table 8.5, there is a considerable difference between the out-of-control ARL values, i.e. the ARL values when $d \neq 0$ of the two charts. Manifestly, the supplementary runs rule increase the sensitivity of the Shewhart chart. However, the supplementary runs rule also reduce the in-control ARL value, i.e. the ARL at the target value μ_0.

Table 8.5
ARL Values for the Two Control Charts

d	C_1	$C_1 \cup C_5$
0.0	370.3983	278.0446
0.2	308.4261	222.5889
0.4	200.0753	134.1700
0.6	119.6653	75.2701
0.8	71.5523	42.9622
1.0	43.8947	25.6122

(Klein 2000) suggested alternatives to the Shewart $\bar{X}$ control chart. Under the scheme defined by (Klein 2000), the process is declared to be out-of-control if two consecutive points both fall on or above the upper control limit (UCL) or both fall on or below the lower control limit (LCL). Let p_U (p_L) denote the probability of a single point falling in the upper (lower) region. Denote by p the probability that a single point falls into the central region, $p = 1 - p_L - p_U$. Define a Markov chain with four states:

State 0: Two consecutive points are beyond just one of the control limits
State 1: No points beyond either control limit
State 2: A point above the upper control limit UCL
State 3: A point below the lower control limit LCL

Cleary, the state 0 is the absorbing state. Thus, the run-length random variable has a phase-type distribution with representation $T \sim PH_3(\pi, \mathbf{Q})$ with $\pi = (1,0,0)$, and

$$\mathbf{Q} = \begin{bmatrix} p & p_U & p_L \\ p & 0 & p_L \\ p & p_U & 0 \end{bmatrix}.$$

The probability mass function of the run-length random variable is obtained as

$$P\{T=n\} = (1,0,0) \begin{bmatrix} p & p_U & p_L \\ p & 0 & p_L \\ p & p_U & 0 \end{bmatrix}^{n-1} \begin{bmatrix} 0 \\ p_U \\ p_L \end{bmatrix},$$

$n = 2,3,\ldots$. The ARL for the chart is

$$ARL = E(T) = (1,0,0) \left(\begin{bmatrix} 1 & 0 & 0 \\ 0 & 1 & 0 \\ 0 & 0 & 1 \end{bmatrix} - \begin{bmatrix} p & p_U & p_L \\ p & 0 & p_L \\ p & p_U & 0 \end{bmatrix} \right)^{-1} \begin{bmatrix} 1 \\ 1 \\ 1 \end{bmatrix}.$$

After some simple manipulations, we obtain

$$ARL = \frac{1}{1 - p - \frac{p_U}{1+p_U} - \frac{p_L}{1+p_L}}.$$

Runs rules have been effectively used to propose better control charts not only parametric but also nonparametric. See, e.g. (Chakraborti et al. 2009), (Koutras and Triantafyllou 2018), (Triantafyllou 2018), and (Chakraborti and Graham 2019) for the evaluation of nonparametric control charts. The second illustration is related to acceptance sampling plan which has been of great interest in statistical quality control. In a conventional setup of the acceptance sampling plan, depending on a certain quality characteristic, a produced item is classified as either nonconforming (defective) or confirming (nondefective). Let X_i denote the status of the ith item such that

$$X_i = \begin{cases} 1, & \text{if the item is nonconforming} \\ 0, & \text{if the item is conforming} \end{cases},$$

$i \geq 1$.

A single sampling plan involves inspecting sequentially n items from the production process submitted for inspection. If the number of nonconforming items $S_n = \sum_{i=1}^{n} X_i$ in the sample is less than or equal to c, then the process is accepted. Otherwise, the process is considered to be out of control. Let AQL and RQL respectively denote the acceptable quality level and rejectable quality level. Then, the sampling plan (n,c) is the plan that satisfy the following requirements:

$$P\{S_n \leq c \mid p_1 = AQL\} \geq 1 - \alpha,$$
$$P\{S_n \leq c \mid p_1 = RQL\} \leq \beta,$$

where α and β respectively denote the producer's and consumer's risk, and $p_1 = P\{X_1 = 1\}$.

(Vellaisamy and Sankar 2001) proposed an algorithm to find single sampling plan with minimum average sample number (ASN). (Vellaisamy and Sankar 2005) obtained single sampling plan under the previous-sum dependent model defined by (7.7) when

$$a_i = (1 - \theta_i)p_1,$$

and

$$b_i = \frac{\theta_i}{i-1},$$

$i \geq 2$, where

$$\max\left(\frac{p_1 - 1}{p_1}, \frac{p_1}{p_1 - 1}\right) \leq \theta_i \leq 1.$$

In this case, the previous-sum dependent model becomes

$$P\{X_i = 1 \mid X_1, X_2, ..., X_{i-1}\} = (1 - \theta_i)p_1 + \frac{\theta_i}{i-1} S_{i-1}. \tag{8.23}$$

Using (7.9), the probability mass function of the number of nonconforming items can be computed from

$$P\{S_n = x\} = ((1-\theta_n)p_1 + \frac{\theta_n}{n-1}(x-1))P\{S_{n-1} = x-1\} \\ + (1-(1-\theta_n)p_1 - \frac{\theta_n}{n-1}x)P\{S_{n-1} = x\}, \tag{8.24}$$

for $n > 1$ and $x = 0, 1, ..., n$ with $P\{S_1 = 1\} = p_1$, $P\{S_1 = 0\} = 1 - p_1$ and $P\{S_n = x\} = 0$ for $x > n$ or $x < 0$.

Example 8.27. For the model (8.23), assume that $\theta_i = 0.1$ for $i \geq 2$. Let $(\alpha, \beta) = (0.01, 0.05)$ and $(AQL, RQL) = (0.01, 0.06)$. Then using the recurrence (8.24), (Vellaisamy and Sankar 2001) computed the optimal plans as $(266, 10)$ with $ASN = 265.69$. ■

Now, consider the extension of acceptance sampling plan to the case when two or more (p, say) quality characteristics are observed on the same produced item. Clearly, since the p characteristics are simultaneously observed on the same item, they are dependent. (Balakrishnan et al. 2009) studied run and frequency quota rules in the context of acceptance sampling plan when each of the p characteristics are classified into three quality levels: "quite close to target quality" "of satisfactory quality" and "far from target quality" which are respectively represented by "0", "1", and "2". Consider the case when $p = 2$. Then, we have a sequence of independent and identically distributed two dimensional random variables $(X_t, Y_t), t = 1, 2, ...,$ where

$$p_{ij} = P\{X_t = i, Y_t = j\},\ i, j = 0, 1, 2,$$

for all $t = 1, 2, ...$

From a large shipment, a lot of size m is taken. Each unit is tested in the lot one by one and classified by one of the three quality levels: "0", "1" or "2". Consider the case when the lot is rejected upon the occurrence of a "2" in either of the two components or k "1"s (in total) in either of the two components. The lot is good and the shipment is accepted if we do not stop by the rule by the time we complete the test on the mth unit. If T_1 represents the waiting time until the first occurrence of the corresponding event, then the probability that the lot is rejected is

$$P\{T_1 \leq m\}.$$

(Balakrishnan et al. 2009) has obtained the following expression for the probability generating function of T_1:

$$\phi(z) = \frac{\sum_{i=1}^{k} b_i z^i}{1 - \sum_{i=1}^{k} (-1)^{i-1} \binom{k}{i} p_{00}^i z^i} = z\pi(\mathbf{I} - z\mathbf{P})^{-1}\mathbf{u}',$$

where $\pi = (1,0,...0)$,

$$\mathbf{P} = \begin{bmatrix} p_{00} & p_{01}+p_{10} & p_{11} & 0 & \cdots & 0 & 0 \\ 0 & p_{00} & p_{01}+p_{10} & p_{11} & \cdots & 0 & 0 \\ 0 & 0 & p_{00} & p_{01}+p_{10} & \cdots & 0 & 0 \\ 0 & 0 & 0 & p_{00} & \cdots & 0 & 0 \\ \vdots & \vdots & \vdots & \vdots & \ddots & \vdots & \vdots \\ 0 & 0 & 0 & 0 & \cdots & p_{00} & p_{01}+p_{10} \\ 0 & 0 & 0 & 0 & \cdots & 0 & p_{00} \end{bmatrix}_{k\times k},$$

and $\mathbf{u}' = (\mathbf{I}-\mathbf{P})\mathbf{e}'$. (Balakrishnan et al. 2009) has also derived the following recursive scheme for the probability mass function of T_1 :

$$P\{T_1 = n\} = \sum_{i=1}^{k}(-1)^{i-1}\binom{k}{i} p_{00}^{i} P\{T_1 = n-i\},$$

for $n \geq k+1$, with initial conditions

$$P\{T_1 = n\} = \left(\sum_{i=0}^{1}\sum_{j=0}^{1} p_{ij}\right)^{n-1}\left(1-\sum_{i=0}^{1}\sum_{j=0}^{1} p_{ij}\right)$$

for $n = 1,2,...,[k/2]-1$ and

$$P\{T_1 = n\} = \sum_{i=1}^{k}(-1)^{i-1}\binom{k}{i} p_{00}^{i} P\{T_1 = n-i\} + b_n$$

for $n = [k/2],[k/2]+1,...,k$.

8.5 START-UP DEMONSTRATION TESTS

As mentioned in the previous sections, a start-up demonstration test is used to elicit information about the reliability of a unit before purchasing it. The unit under test is successively started and the successive outcomes are observed as either a successful start-up or a failure. The decision about the acceptance/rejection of the unit is made based on certain rules related to the occurrences of successes and failures. The first start-up demonstration test called consecutive successes (CS) was introduced and studied by (Hahn and Gage 1983). According to the CS test, the unit is accepted when k consecutive successful start-ups is obtained. In this case, the test length is in fact the number of start-ups until k consecutive successful start-ups. As it was stated in Example 2.10, the test length random variable has geometric distribution of order k when the individual start-ups are independent and identical with the probability of successful start-up p. Define

$$X_i = \begin{cases} 1, & \text{if the outcome of the } i\text{th start-up is success} \\ 0, & \text{if the outcome of the } i\text{th start-up is failure,} \end{cases}$$

for $i = 1,2,\ldots$. Because the tests are applied on the same unit, there might be a dependence between successive test results. We focus on the case when the individual start-ups are Markov dependent with $p_1 = P\{X_1 = 1\}$,

$$p_{01} = P\{X_i = 1 \mid X_{i-1} = 0\},$$
$$p_{11} = P\{X_i = 1 \mid X_{i-1} = 1\},$$

for $i > 1$. Manifestly, the case of independent and identical start-ups is obtained when $p_{11} = p_{01} = p, p_{00} = p_{10} = 1-p, p_1 = p$. For the CS procedure, the test length denoted by T_k has the following probability generating function (Balakrishnan et al. 1995):

$$E(z^{T_k}) = \frac{p_{11}^{k-1} z^k \left[p_1 + (p_{01} - p_1 - p_1 p_{11})z + p_{11}(p_1 - p_{01})z^2\right]}{1 - (p_{11} + 1 - p_{01})z + (p_{11} - p_{01})z^2 + (1 - p_{11})p_{01} p_{11}^{k-1} z^{k+1}}. \quad (8.25)$$

Because the equation (8.25) has the same form with (5.2), the test length random variable T_k has matrix-geometric distribution with probability mass function

$$P\{T_k = t\} = \pi \mathbf{Q}^{t-1} \mathbf{u}',$$

where $\pi = (1,0,\ldots,0)$,

$$\mathbf{Q} = \begin{bmatrix} p_{11} + 1 - p_{01} & 0 & 0 & \cdots & 0 & 1 \\ 0 & 0 & 0 & \cdots & 0 & 0 \\ -(1-p_{11})p_{01}p_{11}^{k-1} & 1 & 0 & \cdots & 0 & 0 \\ 0 & 0 & 1 & & 0 & 0 \\ \vdots & \vdots & \vdots & \ddots & \vdots & \vdots \\ p_{01} - p_{11} & 0 & 0 & & 1 & 0 \end{bmatrix}$$

and

$$\mathbf{u}' = \begin{bmatrix} 0 \\ p_{11}^k (p_1 - p_{01}) \\ p_{11}^{k-1}(p_{01} - p_1 - p_1 p_{11}) \\ p_1 p_{11}^{k-1} \\ \vdots \\ 0 \end{bmatrix}.$$

The CS procedure only considers the successful start-ups and hence the failed start-ups has no effect in decision making process. (Balakrishnan and Chan 2000) proposed a consecutive successes total failures (CSTF) test which is based on both successful and failed start-ups. According to the CSTF procedure, the unit is accepted if k consecutive successful startups are obtained before a total of d failures. An illustration for such a testing procedure has been presented by Example 4.5 when the individual start-ups are independent and identical. Recall that the test length, i.e. the number of start-ups until acceptance/rejection of the equipment can be represented by the waiting time random variable

$$T_{k,d} = \min(T_k, W_d),$$

where T_k is the number of start-ups until k consecutive successful start-ups and W_d is the number of start-ups until d failures. By modifying the structure in Example 4.5, for $k = 4$ and $d = 2$, the distribution of the test length for Markov dependent start-ups can be written as

$$P\{T_{4.2} = t\} = \pi \mathbf{Q}^{t-2}\mathbf{u}',$$

where $\pi = (0, p_1, 0, 0, 1 - p_1, 0, 0, 0)$,

$$\mathbf{Q} = \begin{bmatrix} 0 & p_{11} & 0 & 0 & 1-p_{11} & 0 & 0 & 0 \\ 0 & 0 & p_{11} & 0 & 1-p_{11} & 0 & 0 & 0 \\ 0 & 0 & 0 & p_{11} & 1-p_{11} & 0 & 0 & 0 \\ 0 & 0 & 0 & 0 & 1-p_{11} & 0 & 0 & 0 \\ 0 & 0 & 0 & 0 & 0 & p_{01} & 0 & 0 \\ 0 & 0 & 0 & 0 & 0 & 0 & p_{11} & 0 \\ 0 & 0 & 0 & 0 & 0 & 0 & 0 & p_{11} \\ 0 & 0 & 0 & 0 & 0 & 0 & 0 & 0 \end{bmatrix},$$

and

$$\mathbf{u}' = \begin{bmatrix} 0 \\ 0 \\ 0 \\ p_{11} \\ 1-p_{01} \\ 1-p_{11} \\ 1-p_{11} \\ 1 \end{bmatrix}.$$

In the case of CSTF procedure, the unit under test is either accepted or rejected. The probability of accepting the unit has been derived as (Martin 2004)

$$P\{T_k < W_d\} = 1 - (1 - p_1 p_{11}^{k-1})(1 - p_{01} p_{11}^{k-1})^{d-1}.$$

Example 8.28. Consider the CSTF test when $k = 4$ and $d = 2$. Assume that $p_{11} = 0.9, p_{01} = 0.6$, and $p_1 = 0.9$. Then, the probability of acceptance of the unit is $P\{T_4 < W_2\} = 0.8065$. The mean number of trials until termination of the experiment is found to be $E(T_{4.2}) = 4.3421$. The probability that the test terminates at the tth start-up is given in the following for selected values of t:

$$\begin{aligned} P\{T_{4.2} = 2\} &= 0.040, \\ P\{T_{4.2} = 3\} &= 0.042, \\ P\{T_{4.2} = 4\} &= 0.6993, \\ P\{T_{4.2} = 5\} &= 0.0875. \quad \blacksquare \end{aligned}$$

(Smith and Griffith 2008) defined and studied total successes total failures (TSTF) and consecutive successes consecutive failures (CSCF) tests. According to the TSTF test, the unit is accepted if a total of k successes are observed prior to a total of d failures and rejected otherwise. In the case of CSCF test, the unit is accepted if

k consecutive successful start-ups are observed before d consecutive unsuccessful start-ups. (Eryilmaz 2010c) derived the following expression for the probability mass function of the test length under the TSTF model:

$$P\{Z_{k,d}=t\}=\sum_{m=t-d}^{k-1}p(m,t-1)-\sum_{m=t-d+1}^{k-1}p(m,t),$$

where

$$p(m,t)$$
$$=\sum_{r=1}^{m}\binom{m-1}{r-1}\sum_{i=0}^{1}\sum_{j=0}^{1}\binom{t-m-1}{r-i-j}p_{11}^{m-r}p_{01}^{r-i}(1-p_{11})^{r-j}(1-p_{01})^{t-m-r+i+j-1}p_i,$$

for $0<m<t$, and $t=\min(k,d),\min(k,d)+1,...,k+d-1$, while

$$p(0,t)=p_0(1-p_{01})^{t-1},$$

and

$$p(t,t)=p_1p_{11}^{t-1},$$

where p_1 is the probability of success on the initial start-up and p_0 is the probability of failure on the initial start-up.

Then, the expected test length under the TSTF model can be computed from

$$E(Z_{k,d})=\sum_{t=\min(k,d)}^{k+d-1}tP\{Z_{k,d}=t\}.$$

Example 8.29. Consider the TSTF test when $k=4$ and $d=2$. Assume that $p_{11}=0.9$, $p_{01}=0.6$, and $p_1=0.9$. Then, the mean number of trials until termination of the experiment is found to be $E(Z_{4,2})=4.0967$. The probability mass function of the test length is given in the following:

$$P\{Z_{4,2}=2\}=0.040,$$
$$P\{Z_{4,2}=3\}=0.042,$$
$$P\{Z_{4,2}=4\}=0.6993,$$
$$P\{Z_{4,2}=5\}=0.2187.\quad\blacksquare$$

In Table 8.6, we compute the expected test lengths for TSTF, CSTF, and CSCF models for selected values of transition probabilities, and k and d. Note that the expected test length for the CSCF model is given by (Eryilmaz 2011)

$$\frac{A(p_1p_{10}+p_0p_{01}-p_{10}p_{01})+p_{10}p_{01}+p_1p_{01}(1-p_{11}^{k-1})+p_0p_{10}(1-p_{00}^{d-1})}{p_{10}p_{01}(1-A)},$$

where $A=(1-p_{00}^{d-1})(1-p_{11}^{k-1})$.

Table 8.6
Expected Test Lengths for TSTF, CSTF, and CSCF Models

k	d	p_1	p_{11}	p_{01}	TSTF	CSTF	CSCF
8	5	0.7	0.7	0.5	9.8051	12.1027	35.1312
		0.7	0.9	0.5	9.2116	11.3254	13.4612
		0.9	0.9	0.5	9.0746	11.4091	13.2377
		0.9	0.9	0.6	9.0654	11.7735	13.4653
8	6	0.7	0.7	0.5	10.6814	14.1172	44.7352
		0.7	0.9	0.5	9.5241	12.1298	14.0549
		0.9	0.9	0.5	9.2885	12.0978	13.7460
		0.9	0.9	0.6	9.1988	12.3815	13.7186
10	6	0.7	0.7	0.5	12.3618	15.3696	72.8143
		0.7	0.9	0.5	11.6397	15.3651	19.3992
		0.9	0.9	0.5	11.4730	15.5189	19.1241
		0.9	0.9	0.6	11.4222	16.1238	19.2730
10	8	0.7	0.7	0.5	14.0760	19.8860	107.9917
		0.7	0.9	0.5	12.1118	17.1877	20.2696
		0.9	0.9	0.5	11.8065	17.1478	19.9020
		0.9	0.9	0.6	11.6018	17.5577	19.5584

(Antzoulakos et al. 2009) proposed a modification of the CS type start-up demonstration test. The modified test has been named as CSDF (consecutive successes distance failures). According to the CSDF scheme, if k consecutive successful start-ups are achieved before the appearance of two unsuccessful start-ups separated by at most $r-2$ successful start-ups, then the equipment under test is accepted. If two unsuccessful start-ups separated by at most $r-2$ successful start-ups are achieved before the occurrence of k consecutive successful start-ups, then the equipment under test is rejected. (Antzoulakos et al. 2009) has obtained the following expression for the probability mass function of the test length random variable $W_{k.r}$ when the invidual start-ups are independent and identical with the probability of successful start-up p.

$$P\{W_{k.r}=t\} = \pi\mathbf{Q}^{t-2}(\mathbf{I}-\mathbf{Q})\mathbf{e}',$$

for $t \geq 2$, where

$$\mathbf{Q} = \begin{bmatrix} \mathbf{Q}_{11} & \mathbf{Q}_{12} \\ \mathbf{Q}_{21} & \mathbf{Q}_{22} \end{bmatrix},$$

where

$$\mathbf{Q}_{11} = \begin{bmatrix} 0 & p & 0 & \cdots & 0 \\ 0 & 0 & p & \cdots & 0 \\ \vdots & \vdots & \vdots & \ddots & \vdots \\ 0 & 0 & 0 & \cdots & p \\ 0 & 0 & 0 & \cdots & 0 \end{bmatrix}_{(k-1)\times(k-1)},$$

$$\mathbf{Q}_{12} = \begin{bmatrix} 1-p & 0 & \cdots & 0 \\ 1-p & 0 & \cdots & 0 \\ \vdots & \vdots & \ddots & \vdots \\ 1-p & 0 & \cdots & 0 \end{bmatrix}_{(k-1)\times(r-1)},$$

$$\mathbf{Q}_{22} = \begin{bmatrix} 0 & p & 0 & \cdots & 0 \\ 0 & 0 & p & \cdots & 0 \\ \vdots & \vdots & \vdots & \ddots & \vdots \\ 0 & 0 & 0 & \cdots & p \\ 0 & 0 & 0 & \cdots & 0 \end{bmatrix}_{(r-1)\times(r-1)},$$

and $\mathbf{Q}_{21}$ is is an $(r-1)\times(k-1)$ matrix with all its entries vanishing except for the entry $(r-1,r-1)$ which equals p. (Antzoulakos et al. 2009) has also obtained the following expression for the expected test length:

$$E(W_{k,r}) = \frac{(1-p^k)(2-p^{r-1})}{(1-p)(1+p^k-p^{r-1})}.$$

(Martin 2008) studied TSTF, CSTF, and CSCF models when the underlying binary sequence representing start-up outcomes is Markovian of a general order. The abovementioned start-up demonstration tests have been analyzed via phase-type distributions in (Eryilmaz 2019). There are many different types of start-up demonstration tests that have been studied in the literature. See, e.g. (Gera 2010), (Gera 2011), (Gera 2013), and (Zhao et al. 2015). For a comprehensive review of start-up demonstration tests and their evaluation, see e.g. (Balakrishnan et al. 2014).

REFERENCES

1. Antzoulakos, D. L., Koutras, M.V. and Rakitzis, A.C. 2009. Start-up demonstration tests based on run and scan statistics. *Journal of Quality Technology* 41: 1-12.
2. Balakrishnan, N., Balasubramanian, K. and Viveros, R. 1995. Start-up demonstration tests under correlation and corrective action. *Naval Research Logistics* 42: 1271-1276.
3. Balakrishnan, N., and Chan, P.S. 2000. Start-up demonstration tests with rejection of units upon observing d failures. *Annals of the Institute of Statistical Mathematics* 52: 184-196.
4. Balakrishnan, N., Bersimis, S. and Koutras, M.V. 2009. Run and frequency quota rules in process monitoring and acceptance sampling. *Journal of Quality Technology* 41: 66-81.
5. Balakrishnan, N., Koutras, M.V. and Milienos, F.S. 2014. Start-up demonstration tests: models, methods and applications, with some unifications. *Applied Stochastic Models in Business and Industry* 30: 373-413.
6. Barlow, R.E. and Proschan, F. 1996. *Mathematical Theory of Reliability*, Society for Industrial and Applied Mathematics, Philadelphia.
7. Billinton, R. and Allan, R. 1996. *Reliability evaluation of power systems*, 2nd Edition, Plenum Press, New York.
8. Birnbaum, Z.W. 1969. On the importance of different components in a multicomponent system. *In Multivariate Analysis*, II, Academic Press, New York, 581-592.

9. Chadjiconstantinidis, S. and Koutras, M.V. 1999. Measures of component importance for Markov chain imbeddable reliability structures. Naval Research Logistics 46: 613-639.
10. Chakraborti, S. and Eryilmaz, S. 2007. A nonparametric Shewhart-type signed-rank control chart based on runs. *Communications in Statistics-Simulation and Computation* 36: 335-356.
11. Chakraborti, S., Eryilmaz, S. and Human, S.W. 2009. A phase II nonparametric control chart based on precedence statistics with runs-type signaling rules. *Computational Statistics and Data Analysis* 53: 1054-1065.
12. Chakraborti, S. and Graham, M.A. 2019. Nonparametric (distribution-free) control charts: An updated overview and some results. *Quality Engineering* 31: 523-544.
13. Champ, C.W. and Woodall, W.H. 1987. Exact results for Shewhart control charts with supplementary runs rules. *Technometrics* 29: 393-399.
14. Chang, G.J., Cui, L. and Hwang, F.K. 2000. *Reliabilities of consecutive-k systems*. Kluwer Academic Publishers, The Netherlands.
15. Chien, Y-H. 2012. The effects of renewing free-replacement warranty on the discrete age replacement policy. *IEEE Transactions on Reliability* 61: 389-397.
16. Coolen F.P.A and Coolen-Maturi, T . 2012. Generalizing the signature to systems with multiple types of components. In: Complex systems and dependability 115-130, Berlin: Springer.
17. Coolen F.P.A and Coolen-Maturi, T . 2015. Modelling uncertain aspects of system dependability with survival signatures. In: Dependability problems of complex information systems 19-34, Berlin: Springer.
18. Da, G., Zheng, B. and Hu, T. 2012. On computing signatures of coherent systems. *Journal of Multivariate Analysis* 103: 142-150.
19. Davies, K. and Dembinska, A. 2019. On the number of failed components in a k-out-of-n system upon system failure when the lifetimes are discretely distributed. *Reliability Engineering and System Safety* 188: 47-61.
20. Dembinska, A. and Eryilmaz, S. 2021. Discrete time series-parallel system and its optimal configuration. *Reliability Engineering and System Safety* 215: 107832.
21. Eryilmaz, S. 2010a. Review of recent advances in reliability of consecutive k-out-of-n and related systems. *Proceedings of the Institution of Mechanical Engineers, Journal of Risk and Reliability* 224: 225-237.
22. Eryilmaz, S. 2010b. Mixture representations for the reliability of consecutive-k systems. *Mathematical and Computer Modelling* 51: 405-412.
23. Eryilmaz, S. 2010c. Start-up demonstration tests under Markov dependence. *Pakistan Journal of Statistics* 26: 637-647.
24. Eryilmaz, S. 2011. Joint distribution of run statistics in partially exchangeable processes. *Statistics & Probability Letters* 81: 163-168.
25. Eryilmaz, S. 2012. The number of failed components in a coherent system with exchangeable components. *IEEE Transactions on Reliability* 61: 203-207.
26. Eryilmaz, S. 2014. On signatures of series and parallel systems consisting of modules with arbitrary structures. *Communications in Statistics - Simulation and Computation* 43, 1202-1211.
27. Eryilmaz, S. 2016. Consecutive k-out-of-n lines with a change point. *Proceedings of the Institution of Mechanical Engineers, Journal of Risk and Reliability* 230: 545-550.
28. Eryilmaz, S. 2019. Statistical inference for a class of startup demonstration tests. *Journal of Quality Technology* 51: 314-324.

29. Eryilmaz, S. 2021. Revisiting discrete time age replacement policy for phase-type lifetime distributions. *European Journal of Operational Research* 295: 699-704.
30. Eryilmaz, S. and Kan, C. 2021a. Reliability assessment for discrete time shock models via phase-type distributions. *Applied Stochastic Models in Business and Industry* 37: 513-524.
31. Eryilmaz, S. and Kan, C. 2021b. A new shock model with a change in shock size distribution. *Probability in the Engineering and Informational Sciences* 35: 381-395.
32. Eryilmaz, S. and Sarikaya, K. 2014. Modeling and analysis of weighted-k-out-of-n:G system consisting of two different types of components. *Proceedings of the Institution of Mechanical Engineers, Journal of Risk and Reliability* 228: 265-271.
33. Eryilmaz, S. and Tank, F. 2021. Computing minimal signature of coherent systems through matrix-geometric distributions. *Journal of Applied Probability* 58: 621-636.
34. Gera, A. E. 2010. A new start-up demonstration test. *IEEE Transactions on Reliability* 59: 128-131.
35. Gera, A. E. 2011. A general model for start-up demonstration tests. *IEEE Transactions on Reliability* 60: 295-304.
36. Gera, A. E. 2013. A start-up demonstration procedure involving dependent tests. *Statistics & Probability Letters* 83: 2191-2196.
37. Gertsbakh, I., Shpungin, Y. and Spizzichino, F. 2011. Signatures of coherent systems built with separate modules. *Journal of Applied Probability* 48: 843-855.
38. Gut, A., and Hüsler, J. 1999. Extreme shock models. *Extremes* 2: 293-305.
39. Hahn, G. J. and Gage, J. B. 1983. Evaluation of a start-up demonstration test. *Journal of Quality Technology* 15: 103-106.
40. Kan, C. and Eryilmaz, S. 2021. Reliability assessment of a discrete time cold standby repairable system, *TOP* 29: 613-628.
41. Klein, M. 2000. Two alternatives to the Shewhart $\bar{X}$ control chart. *Journal of Quality Technology* 32: 427-43
42. Koutras, M.V. 1996. On a Markov chain approach for the study of reliability structures. *Journal of Applied Probability* 33: 357-367.
43. Koutras, M.V. and Triantafyllou, I. 2018. A general class of nonparametric control charts. *Quality and Reliability Engineering International* 34: 427-435.
44. Kuo, W. and Zuo, M. 2003. *Optimal Reliability Modeling, Principles and Applications*, John Wiley and Sons, New Jersey.
45. Kuo, W. and Zhu, X. 2012. *Importance Measures in Reliability, Risk and Optimization*, John Wiley and Sons, United Kingdom.
46. Li, Z. H. and Kong, X.B. 2007. Life behavior of δ-shock model. *Statistics and Probability Letters* 77: 577-87.
47. Li, W. and Zuo, M.J. 2008. Reliability evaluation of multi-state weighted k-out-of-n systems. *Reliability Engineering and System Safety* 93: 160-167.
48. Malinowski, J. and Preuss, W. 1995. A recursive algorithm evaluating the exact reliability of a consecutive-k-within-m-out-of-n:F system. *Microelectronics and Reliability* 35: 1461-1465.
49. Mallor, F. and Omey, E. 2001. Shocks, runs and random sums. *Journal of Applied Probability* 38: 438-448.
50. Martin, D.E.K. 2004. Markovian start-up demonstration tests with rejection of units upon observing d failures. *European Journal of Operational Research* 155: 474-486.
51. Martin, D.E.K. 2008. Application of auxiliary Markov chains to start-up demonstration tests. *European Journal of Operational Research* 184: 574-583.

52. Navarro, J., Ruiz, J. M. and Sandoval, C. J. 2007. Properties of coherent systems with dependent components. *Communications in Statistics-Theory and Methods* 36:175-191.
53. Navarro, J. and Rubio, R. 2010. Computations of signatures of coherent systems with five components. Communications in Statistics- Simulation and Computation 39: 68-84.
54. Nakagawa, T., Osaki, S. 1977. Discrete time age replacement policies. *Operational Research Quarterly*, 28: 881-885.
55. Nakagawa, T. 1985. Continuous and discrete age replacement policies. *Journal of the Operational Research Society* 36: 147-154.
56. Papastavridis, S. 1987. The most important component in a consecutive k-out-of-n:F system. *IEEE Transactions on Reliability* 36: 266-268.
57. Papastavridis, S. 1989. The number of failed components in a consecutive k-out-of-n:F system. *IEEE Transactions on Reliability* 38: 338-340.
58. Parvardeh, A. and Balakrishnan, N. 2015. On mixed δ-shock models. *Statistics and Probability Letters* 102: 51-60.
59. Samaniego, F. J. 2007. *System signatures and their applications in engineering reliability*. Springer, New York.
60. Smith, M. L., and Griffith, W.S. 2008. The analysis and comparison of start-up demonstration tests. *European Journal of Operational Research* 186: 1029-1045.
61. Tong, Y.L. 1985. A rearrangement inequality for the longest run with an application to network reliability. *Journal of Applied Probability* 22: 386-393.
62. Triantafyllou, I. 2018. Nonparametric control charts based on order statistics: Some advances. *Communications in Statistics-Simulation and Computation* 47: 2684-2702.
63. Vellaisamy, P. and Sankar, S. 2001. Sequential and systematic sampling plans for Markov dependent production process. *Naval Research Logistics* 48: 451-467.
64. Vellaisamy, P. and Sankar, S. 2005. A unified approach for modeling and designing attribute sampling plans for monitoring dependent production processes. *Methodology and Computing in Applied Probability* 7: 307-323.
65. Wang, G. J. and Zhang, Y.L. 2005. A shock model with two-type failures and optimal replacement policy. *International Journal of Systems Science* 36: 209-214.
66. Wu J.S. and Chen R.J. 1994. An algorithm for computing the reliability of a weighted-k-out-of-n system. *IEEE Transactions on Reliability* 43: 327-328.
67. Zhao, X., Wang, X. and Sun, G. 2015. Start-up demonstration tests with sparse connection. *European Journal of Operational Research* 243: 865-873.

Index

Made in the USA
Las Vegas, NV
19 September 2026

7d949254-05fe-4844-b8a3-bb5065625e94R01